Digital Self Defense

Also from EATMS Productions

Books on power, survival, women's autonomy, and the systems shaping modern America.

Nonfiction

Billionaires, Capitalism, and Power

Evil and the Mountain Ungreed
Self Help for American Billionaires
Selfish Steve and the Ivory Tower
Tariffs, Taxes, & Face-Eating Leopards
Ban Billionaires: Fascism Fix

Fascism, Religion, and Cultural Control

Self Help for the Manosphere
Fascism 2025
Fascism & the Perverts & the Greed Virus
Christian Fascism Marriage Book
Tyranny, Table Manners, & Tiramisu

Guides for Women's Autonomy and Protection

How to Survive in Post-America as a Woman
Project 2025 American Drag
4B – Burn, Ban, Boycott, Build
4B OG – So No Go GYN
I'm Glad He's Dead

Analysis of Authoritarian Project 2025

Project 2025: The Blueprint
Project 2025: The List
Project 2025, Christian Dumb Dumbs, & The Republican Agenda
Fascism, Project 2025, & The Pinkprint

Modern Rewrites for Women

Stoic Principles Reimagined
Siddhartha Reimagined
The Prince Reimagined for Women
The Art of War Reimagined for Women
The Jungle Reimagined
The Constitution Reimagined for Women

Machine Learning Series

AI, Bitcoin, Nostr for Women
AI, Safety, & Security for Women
AI, Anxiety, & Health for Women
AI, Kids, & Family Safety for Women
AI, Creativity, & Personal Expression for Women
AI, Independent Work, & Parallel Power for Women

Social Systems Series

Emotional Labor for Women
Household Power for Women
Workplace Power for Women
Medical Bias for Women
Aging Systems for Women
Recovery Systems for Women

Fiction

Dystopian Stories of Resistance and Collapse

Propaganda Paige & the Missing Prosperity
Propaganda Paige & the TIDE Manifesto
Propaganda Paige & the Shadow Cartographers
Propaganda Paige & the Prosperity Alliance
Propaganda Paige & the Shattered Truth
Propaganda Paige & the Rising TIDE
Propaganda Paige & the Last Bastion
Propaganda Paige & the Dawn of Prosperity
Project 2025: Dorian — The Last Men
Project 2025: Boy — A Last Men Novel

Digital Self Defense for Women

The Autonomy Series
Book 3

by
Mads Duchamp

EATMS
PRODUCTIONS

ISBN: 978-1-966014-47-8

Cover, interior design by: Esme Mees

eatms@pm.me
www.eatms.me

Check out EATMS Underground:
https://tinyurl.com/eatmsNOSTR

Printed in the United States of America.

Surveillance does not need hostility to function. It needs routine.

— Hannah Arendt

Table of Contents

Foreword

Digital life shapes women's choices long before anyone explains what is happening. It determines what she can access, what she can't undo, how visible she becomes without consent, and how easily systems can be used against her when conditions tighten. Women are told to treat technology as convenience, a tool that works in the background. Tracking is framed as personalization. Monitoring is framed as safety. Over time, that confusion becomes a quiet form of control, especially in environments where compliance is rewarded and refusal is punished without explanation.

Digital self-defense enters this landscape not as paranoia, but as clarity. It exposes how much of modern life is built on surveillance, extraction, and exposure. For women, those defaults matter. A password leak is not inconvenience. A device search is not policy. A data broker is not advertising. These systems create leverage. They create records that outlast context. They create friction for women trying to move, leave, protect children, protect work, or protect privacy under pressure.

This book focuses on understanding that shift. It examines how digital systems classify people, how tracking becomes power, and why women face higher consequences when information becomes weaponized. The goal is not to turn women into technicians. It is to make the structure visible so choices are deliberate rather than assumed.

—Mads Duchamp, Winter 2026

~ 1
Living Inside Invisible Systems

The Constraint

Women are told the digital world is neutral. Apps, platforms, devices, and services are framed as convenience tools. You open your phone, do what you need, and move on. But modern life is not built on tools. It is built on systems. Systems decide what becomes easy, what becomes exhausting, what is allowed, and what requires permission. Women do not simply use technology. They live inside it.

An invisible system is anything that shapes behavior without needing your consent each time. It is the default login method. It is the requirement to verify identity. It is the way an app demands location access to function. It is the "security" update that changes your settings and calls it improvement. It is the automated decision that blocks your purchase and offers no human path to resolve it. These mechanisms are not rare. They are routine. They are built into the architecture of daily life.

Most of the power does not come from obvious surveillance. It comes from quiet dependency. When every part of life is mediated through accounts, a woman becomes dependent on systems she cannot negotiate with. If your phone number is the master key to your identity, losing it can lock you out of banking, email, medical portals, work tools, school communications, and family coordination. The system does not care if you need access urgently. It only recognizes whether you can satisfy the required steps. When you cannot, your life pauses.

This dependency is intensified by consolidation. Women are often encouraged to keep everything under one "ecosystem." One email. One cloud account. One messaging app. One password reset method. One device. This is marketed as simplicity. In practice, it becomes fragility. One failure point becomes a cascade. The woman loses access not because she made a mistake, but because the system was designed to collapse fast when something looks abnormal.

These systems also reward compliance and punish friction. If you give the app what it wants, the process is smooth. If you refuse, it becomes inconvenient. The screen repeats the request. The button to continue becomes hard to find. The app warns you that features will not work. This is not neutral design. It is pressure. It trains behavior by making the safest choices feel like the hardest choices.

Women absorb this pressure in unique ways because women often manage the practical infrastructure of other people's lives. Care coordination. Scheduling. School contact. Medical follow-ups. Travel documents. Password resets for relatives. Bills. Family logistics. When systems break, women do the repair labor. They sit on hold. They resend forms. They navigate appeals. They stay calm while everyone else is blocked. The system calls this user experience. For women, it becomes unpaid labor under digital constraint.

The invisible system also shapes social safety. Women are pushed into constant visibility. Platforms are built to encourage sharing, linking, tagging, and location stamping. The modern expectation is that everyone is reachable and searchable. Women who opt out are treated as strange or suspicious. But visibility is not always connection. Often, it is exposure. It turns a normal life into a searchable profile that strangers, employers, partners, and institutions can interpret however they want.

This becomes more dangerous in periods of instability. A woman may need to leave quickly. She may need to change routines. She may need to reduce contact. She may need to protect children. She may need to move money quietly. But systems do not care about context. They enforce policies. They log actions. They trigger alerts. They mark behavior as suspicious when a woman acts outside her typical pattern. A protective move becomes a risk marker. A survival choice becomes a red flag.

In centralized environments, the consequences are easier to predict. There are policies and appeals, even if they are weak. In decentralized or loosely governed environments, the consequences are less predictable. But in both cases, the pattern is the same: the system decides. The woman reacts. The woman adapts. The woman pays the cost.

A digital system does not have to be explicitly authoritarian to create control. It only needs to require permission, record behavior, and enforce rules through automated gates. This is how modern life becomes

governable without looking like governance.
Restrictions do not need to be announced. They can be
embedded as defaults. They can be framed as safety.
They can be introduced as convenience. And once the
infrastructure exists, it can be used in ways the original
design never stated.

Women are often told that if they are careful, they will
be fine. But care is not the real issue. The structure is. A
woman can be thoughtful, cautious, competent, and still
lose access because a system changed its rules. She can
still be exposed because a platform pushed her
information outward. She can still be punished because
an algorithm misread her actions. When systems hold
the power to decide access, the question is not whether
women make perfect choices. The question is how
fragile the environment is when pressure arrives.

This is the constraint: digital life is structured as
invisible governance. It trains behavior through friction.
It creates dependency through consolidation. It
increases exposure through default visibility. It collapses
autonomy through single points of failure. Women can
feel the pressure without being able to name it, because
the system does not present itself as control. It presents
itself as the only way to function.

The constraint becomes sharper when women
recognize how many decisions are no longer choices. A
woman does not choose to create ten accounts. She is
required to. She accepts defaults because refusing
means losing access. She does not choose to store
private life in corporate clouds. She is routed there by
design. Participation is assumed, and the penalty for

refusal is inconvenience, isolation, and loss of basic services.

Many women do not realize how quickly informal norms become enforced norms. A workplace begins with optional authentication and later makes it mandatory. A school begins with email notices and later requires an app. A doctor's office begins with phone calls and later routes everything through a portal that logs every message. Women adapt gradually, then notice what was surrendered only when they try to pull back.

These systems do not just govern access. They govern emotional energy. They create constant low-level vigilance: managing settings, updating passwords, confirming identity, approving logins, and chasing down errors that were not her fault. That cost is invisible, but it accumulates into exhaustion and quiet compliance.

This is the real shape of invisible systems: they make women legible to institutions while making women's context irrelevant. They make compliance feel effortless and resistance feel socially costly. And they normalize a world where a woman's freedom depends on satisfying processes that were not built for her protection.

The Shift

The shift begins when a woman stops treating digital life as a set of personal habits and starts treating it as infrastructure. Infrastructure is not a vibe. It is not a relationship. It is not a reflection of your values. It is built to do specific things at scale: identify users, route access, store records, and enforce rules. Once you see that clearly, the shame and confusion fall away. You stop asking why the system feels suffocating and start noticing how it works. You stop trying to earn comfort from a structure that was never designed to provide it.

This is not a shift into paranoia. It is a shift into accuracy. Invisible systems feel personal because they sit inside your daily life. They ask for your attention constantly. They create problems you must fix. They punish you for mistakes you did not make. But the system does not know you. It does not care about your context. It does not recognize your intentions. It recognizes inputs. It recognizes patterns. It recognizes compliance.

Once women accept this, the task becomes simpler. The task is not to be understood by the system. The task is to build a life that remains functional even when the system does what it always does.

The first change is learning to spot coercion disguised as convenience. Many systems are built with a quiet threat underneath the "free" or "easy" offer. If you do this now, it will be smooth. If you refuse, it will become difficult. This is how platforms train behavior. They do not need to argue with women. They only need to make the safest choice feel inconvenient. A woman who notices this is already regaining leverage. She can pause, name the pressure, and decide deliberately instead of being pushed forward by friction.

The second change is separating identity from access. Most modern systems try to collapse everything into one master key. Your phone number becomes your identity. Your email becomes your identity. Your cloud account becomes your identity. Once that happens, one disruption cascades into everything. The shift is recognizing that a woman's life should not hinge on one credential. She does not need to become a security professional. She only needs to stop building her daily functioning on single points of failure.

This is where digital self-defense becomes practical. It is not about being invisible. It is about being resilient. Resilience means redundancy. It means a backup method of access. It means a second route to recovery. It means the ability to keep going even when a system locks her out, changes its rules, or misreads her behavior. Most women already understand redundancy in real life. They keep extra keys. They keep spare supplies. They keep emergency plans for children and work. Digital life deserves the same seriousness because digital life now controls the doors.

Another part of the shift is learning what actually matters. Women get overwhelmed because "digital safety" is marketed as endless threat headlines and endless products. That framing is meant to keep people anxious and dependent. The real structure is smaller.

There are a handful of critical access points that determine most outcomes: your primary email, your phone number, your banking access, your device unlock, your password storage, and your cloud storage. When those are stable, everything else becomes manageable. When those are fragile, everything else becomes chaos.

A woman can also reclaim power by limiting unnecessary exposure. The system's default posture is to collect, store, and share. That is the business model. The shift is choosing minimization instead. Not because you have something to hide, but because you have something to protect. Your time. Your location. Your relationships. Your children. Your work. Your ability to move without being tracked. Minimization is not about withdrawing from life. It is about reducing the amount of leverage the system can accumulate against you.

This shift is psychological as much as it is technical. Many women have been trained to interpret digital inconvenience as personal failure. A password reset fails and she feels incompetent. An account gets flagged and she feels ashamed. A device behaves unpredictably and she feels behind. But the system is intentionally complex. It is designed to be navigated by compliance, not understanding. Once women let go of the idea that competence means total control, they can redirect energy toward the only thing that matters: reducing fragility and increasing options.

Women also have to stop expecting fairness from systems that are optimized for efficiency. A system that can instantly lock an account cannot instantly understand context. A system that can track everything cannot guarantee protection. A system that can enforce rules automatically cannot offer nuanced judgment. The shift is recognizing that the most dangerous moment is the moment you expect the system to "be reasonable." Systems do not become reasonable under stress. They become stricter, colder, and more automated.

This matters because women often face safety needs under stress. Leaving. Relocating. Changing routines. Ending contact. Moving money. Protecting children. Protecting work. These moments demand privacy, speed, and control. But systems interpret those actions as suspicious. Sudden changes trigger flags. New devices trigger verification. New locations trigger alerts. A woman's attempt to protect herself can produce the exact pattern the system is built to challenge.

The shift also includes accepting that some tradeoffs are permanent. Convenience has a cost. Centralization has a cost. Being reachable has a cost. Storing life in platforms has a cost. Women do not need to reject everything. But they do need to stop pretending the costs are imaginary. The most stable posture is intentional compromise. A woman decides what she will expose, and what she will not. She decides which systems are allowed into the center of her life and which are kept at the edges.

This shift changes how women evaluate tools. The question is not "Is this app popular?" It is "What dependency does this create?" The question is not "Will this make things easier today?" It is "Will this make leaving harder later?" The question is not "Is this safe?" It is "How does failure look, and can I recover quickly?" When women ask these questions, they stop being managed by invisible systems and start managing their relationship to them.

Women do not need to win a war against technology. They need to build autonomy inside it. Autonomy is not a fantasy where no one tracks you and no institution has power. Autonomy is the ability to make choices under pressure without collapsing. It is the ability to move, recover, and rebuild without begging for permission from systems that do not care why you need it.

This chapter is about seeing the environment correctly. Invisible systems shape daily decisions because they control access, log behavior, and punish resistance through friction. The shift is refusing to confuse that structure with normal life. Once a woman sees the system as infrastructure, she can build redundancy, minimize exposure, and reduce single points of failure. She can keep convenience where it helps and protect autonomy where it matters.

The goal is not fear. The goal is options. When systems tighten, women with options can adapt. When systems fail, women with redundancy can recover. When systems misread behavior, women with multiple routes can keep moving. The shift is learning to live inside invisible systems without letting them become the author of your life.

~2
Data as a Form of Power

The Constraint

Women are told data is harmless. Apps and platforms call it personalization. Devices call it convenience. Websites call it "improving your experience." A woman is taught to believe that data is just the price of modern life. Click accept. Keep moving. But modern life is not built on harmless collection. It is built on extraction. Systems do not collect because they care about you. They collect because the record becomes power.

A data point is anything that can be stored and reused. It is where you go. It is who you speak to. It is what you search when you are anxious. It is what you buy when you are trying to cope. It is what time you wake up, what time you fall asleep, what time you stop replying. It is your device identifier, your location history, your payment methods, and your app permissions. Data is not one file. It is a web of signals that accumulates quietly.

Most of the power does not come from one dramatic leak. It comes from accumulation. A single location ping means little. A month of location pings becomes a routine. A year of purchases becomes a pattern. A long message history becomes a map of your relationships, your stress, your support, and your dependency. Women do not notice this process because each moment feels small. But systems are designed to make the small moments permanent.

Women are taught the wrong privacy question. They are told to ask whether they have something to hide. That question frames privacy as guilt. It makes boundaries sound suspicious. But privacy is not about hiding wrongdoing. It is about controlling access. It is about limiting who can build leverage over your life. A woman can be doing nothing wrong and still be exposed, profiled, tracked, and targeted.

Data creates power because it reduces uncertainty for whoever holds it. It allows prediction. It allows influence. It allows restriction. It lets companies guess what will move your attention. It lets institutions decide whether you are worth serving or worth denying. It gives employers shortcuts. It gives insurers risk models. It gives governments infrastructure for quiet enforcement. Data turns a person into a pattern that can be managed without conversation.

For women, the consequences are sharper because women live inside higher stakes vulnerability. Women are more likely to be punished socially for the same behavior. More likely to be blamed when something goes wrong. More likely to be evaluated through rumor, implication, and "context" that no one verifies. Data does not need to be correct to be dangerous. It only needs to be believable. And once the record exists, women have to live with the interpretations others build on top of it.

Women's data is also more valuable because women are often the operational center of daily life. A woman's device is not just personal. It holds household logistics, children's schedules, school communications, medical reminders, caregiving coordination, and work planning. Her accounts become the pathways through which other people's stability runs. When her phone is tracked, the family is trackable. When her accounts are profiled, the household becomes legible. The system is not only collecting a woman. It is collecting the structure around her.

A woman's calendar is not just a schedule. It is a forecast of where she will be and when. It can reveal predictable locations and predictable gaps. It can expose routines without the woman ever "posting" anything publicly. A woman's contacts are not just names. They are a map of the people she can reach when something breaks. A woman's photos are not just memories. They contain faces, documents, addresses, school logos, license plates, and background clues that can be extracted later.

This is why data is more valuable than women assume. It is dense and relational. It includes the invisible labor of life management. It includes children. It includes partners. It includes parents. It includes clients, workplaces, appointments, and routines. A woman's phone often contains the truth of her life more clearly than any social profile. That truth can be used for marketing. It can also be used for pressure.

Data also outlasts context. A woman can leave a relationship and assume it is over. She can move cities and assume the old life is behind her. She can change jobs, change routines, and change her mind. But the record stays. Old accounts persist. Old purchases persist. Old messages persist. Old location patterns can be reconstructed. Data systems do not forget because forgetting is not profitable.

Once a record exists, it becomes portable. It can be copied, sold, merged, and reused without a woman's awareness. Data brokers compile profiles from public records, shopping behavior, web activity, and app data. Advertisers trade identifiers that follow women across devices and across sites. "Anonymous" datasets become identifiable when enough pieces are combined. The system does not need your name to know who you are. It only needs enough consistency to lock onto you.

Women often experience the consequences indirectly. Ads appear at moments that feel invasive. Recommendations arrive that feel too accurate. Pricing shifts quietly. Content funnels tighten. A woman is shown a version of the world that pushes her toward a decision she did not consciously choose. None of this requires a villain. It requires an incentive. The incentive is prediction, and prediction requires data.

The constraint becomes more serious when data is used in enforcement systems. Fraud detection. Identity verification. Account risk scoring. Content moderation. Eligibility checks. These systems operate through pattern matching, not understanding. They do not know your story. They only know whether your behavior looks like what they expect. Women are more likely to trigger friction because women's lives are more likely to shift abruptly under pressure. New jobs. New homes. New routines. Shared accounts. Emergency purchases. A move that looks suspicious to a system can be normal to a woman trying to survive.

This is how data becomes control without calling itself control. A system delays a transfer. It forces extra verification. It locks an account. It blocks access. It demands documentation. It asks for proof that a woman may not be able to produce quickly. The system calls this security. The woman experiences it as restriction. And when she is responsible for keeping life running, restriction becomes panic.

Women are told that data collection is the cost of participation. They are told that tracking is normal and inevitable. They are told that the tools are free, so they should stop complaining. But the cost is not just "ads." The cost is exposure that cannot be reversed. The cost is a record that can be used in ways a woman never agreed to. The cost is a system that knows her patterns better than she realized, and can act on that knowledge without needing her permission.

This is the constraint: data is power held by others. It accumulates quietly. It travels without consent. It outlasts context. It becomes enforcement when pressure rises. And women carry the cost first, because women live closer to consequence.

The Shift

The shift begins when a woman stops treating data as a side effect of modern life and starts treating it as power. Not her power. Someone else's. Data is the ability to predict, target, pressure, and restrict without asking permission. It is the ability to map your routines, your relationships, and your weak points, then act on that map through systems that never have to explain themselves. Once a woman understands that, she stops asking whether privacy is "worth it" and starts asking how much leverage she wants other people to hold.

Women are trained to think privacy is about secrecy. That framing is dead on arrival. Most women are not hiding wrongdoing. They are trying to live without being exposed, profiled, or cornered. The shift is recognizing that privacy is not a moral category. It is a boundary. It is a decision about what gets stored, what gets merged, and what gets reused when conditions tighten. A woman can be honest and still require limits. She can be social and still refuse unnecessary collection.

The most important shift is dropping the "nothing to hide" trap. That line turns privacy into guilt and makes women feel unreasonable for wanting control. But data risk is not about guilt. It is about inference. It is about how easy it becomes for strangers, institutions, and systems to decide who you are based on partial records. A woman does not need to be doing anything wrong to be harmed. She only needs to be legible enough for someone else to act.

Once women see that, the problem becomes simpler. The main exposure comes from defaults. Apps ask for location even when they do not need it. Platforms ask for contacts even when they do not need them. Devices request "always-on" permissions because constant collection is more valuable than occasional collection. The screen makes "Allow" fast and "Don't Allow" feel like refusal. This is not an accident. It is pressure disguised as convenience.

The shift is learning to treat those requests as negotiations, not instructions. A permission pop-up is a bargaining moment. It is a system trying to deepen the profile it has on you. Women are allowed to say no. They are allowed to select "only while using." They are allowed to deny access without apologizing to a screen. This does not require perfection. It requires a new default posture: minimize inputs, especially the ones that cannot be taken back.

Women also need to stop treating "data" as one abstract blob. A handful of categories create most of the leverage. Location history. Messages. Photos. Contact lists. Password recovery methods. Financial accounts. These are not theoretical. They are the practical map of your life. They describe where you go, who you rely on, who relies on you, what you keep, and how someone could interrupt your stability. Protecting these is not a niche concern. It is basic resilience.

The shift also includes recognizing that permanence is the real danger. A woman can handle inconvenience today. The problem is a record that follows her for years. Systems store because storage is cheap and valuable. They retain because retention increases prediction. They share because sharing increases monetization. A woman does not have to accept permanent archive as the default condition of her life. She can reduce what exists, reduce how long it exists, and reduce how widely it spreads.

This is where women regain control through separation. Not withdrawal. Separation. Compartmentalization is how strong systems reduce damage. When everything is linked, one login, one cloud, one phone number, one ecosystem, one failure becomes a collapse. A woman doesn't need to build a fortress. She needs to stop building her life around single points of failure. The shift is creating distance between the parts of life that should never fall together in one breach.

Convenience will try to seduce women back into fragility. Auto-fill. Cross-device syncing. Smart suggestions. "Memories." All of it is sold as relief. Some of it is useful. But the cost is deeper profiling and deeper dependency. The shift is choosing convenience selectively instead of automatically. A woman decides where convenience is worth the trade and where it is not, based on consequence instead of habit.

Another shift is expanding the threat model beyond "hackers." Theft is real, but it is not the only risk. The more common risk is institutional use. Employers filtering candidates. Insurers pricing risk. Platforms moderating and shadowbanning. Banks freezing transfers. Schools logging communications. Healthcare portals storing sensitive history. Data brokers compiling and reselling. None of these systems need bad intent to create harm. They only need incentives that prioritize the institution over the individual.

Women feel this most when automated enforcement collides with real life. Fraud detection does not understand emergencies. Identity verification does not care why a woman moved. Risk scoring does not pause for context. These systems enforce patterns, not stories. They treat deviation as danger and demand proof when a woman is least able to provide it. The shift is preparing for that friction ahead of time, instead of meeting it unarmed in the moment.

Preparation looks boring, which is why it works. A woman strengthens the few access points that control everything else. She makes sure her core accounts are recoverable. She reduces the number of places that can lock her out of her own life. She keeps copies of critical information offline or in a separate place. She makes sure that losing one device does not mean losing everything. The shift is building redundancy so survival does not depend on one fragile account behaving perfectly.

Women also have to let go of the fantasy of fairness. Systems are not built to protect women. They are built to reduce liability and increase compliance. A woman who expects a system to "be reasonable" under pressure is likely to be disappointed. The shift is expecting that friction will happen, and structuring life so friction is survivable. This is not pessimism. It is accurate. It is how women keep moving when the system tightens.

This chapter is about changing what "safety" means in the data age. The constraint is that data becomes leverage because it accumulates, spreads, and outlasts context. It turns women into profiles that can be targeted and restricted without conversation. The shift is not disappearance. It is deliberate minimization, separation, and redundancy. Women reduce what can be collected, reduce what can be stored permanently, and reduce what can be merged into a single point of control.

The point is not to be invisible. The point is to be harder to corner. A woman who reduces the record reduces leverage. A woman who separates categories of life reduces blast radius. A woman who builds redundancy keeps access when systems misread her. That is what digital self-defense looks like at the level that actually matters: not perfect privacy, but durable options under pressure.

~3
Surveillance as the Default

The Constraint

Many women think surveillance is something dramatic.
A camera on a pole. A government database. A stalker
watching a house. Something obvious. Something
illegal. Something that happens to other people. But
modern surveillance is not an event. It is an
environment. It is the default condition of daily life,
built into phones, apps, cars, homes, workplaces, and
schools. Women do not step into surveillance by
making one bad choice. They live inside it because the
system is designed that way.

Surveillance became normal by being marketed as
convenience. Location services make life easier. Fitness
tracking makes health feel measurable. Smart devices
make homes feel efficient. Autofill makes logins fast.
Sync makes everything seamless. Women are
encouraged to accept these features because they
reduce friction. But the price of seamlessness is
continuous measurement. The system learns your
habits because it is always watching your inputs.

Most tracking is not framed as tracking. It is framed as
product improvement. "Diagnostics." "Analytics."
"Personalization." "Safety." "Fraud prevention." Each
of these labels gives surveillance a friendly face. They
make it sound responsible and protective. But the
function is the same. Data is collected. Data is stored.

Data is shared. A record is created that outlives the moment it was generated.

Women are also trained to consent without noticing. The screen flashes a request. Accept to continue. Allow to proceed. Agree to use the service. Refusing requires effort. It requires time. It requires reading. It requires knowing which settings matter. It requires pushing against social pressure that says you are being difficult. So women accept. Not because they want surveillance. Because they want to finish the task and get on with life.

Once consent becomes routine, surveillance becomes invisible. It is not one permission. It is a thousand small permissions stacked into normal life. A woman allows location "just this once" for weather and never checks again. She allows microphone access for one voice note and forgets. She allows photo access for one upload and now the app sees everything. She allows contacts for convenience and now her network is mapped. Most women are not signing away privacy in one moment. They are bleeding it out over time.

The constraint is deeper because surveillance is not just about what you share. It is about what is inferred. A woman does not have to post her location publicly for systems to know where she is. Her phone pings towers. Apps read IP addresses. Devices share identifiers. Stores track movement through Wi-Fi and Bluetooth. Cars record routes. Cameras read license plates. A woman can live quietly and still be tracked because the environment is designed to observe by default.

This surveillance is not limited to big tech platforms. It extends into workplaces, schools, and medical systems. Work devices are monitored. Email is logged. Browsers are tracked. Security tools record clicks. School portals track parent activity. Apps record attendance and communication. Healthcare portals store sensitive history. None of this is presented as surveillance. It is presented as administration. Women are expected to accept it because it is "how things work."

Surveillance also becomes social. Platforms encourage constant visibility and constant reachability. Read receipts. Online status. Last active. Location sharing. Tagging. Automatic memories. A woman is nudged to document and confirm her existence continuously. If she turns features off, she is treated as suspicious or cold. If she limits access, she is treated as untrusting. Surveillance becomes enforced not only by technology, but by social expectation.

Women pay a higher cost because women face a higher threat surface. Women are more likely to be stalked. More likely to be harassed. More likely to be punished for refusing attention. Tracking makes those threats easier. It turns a woman's routines into a schedule that can be exploited. It turns her relationships into a network that can be pressured. It turns her moments of vulnerability into predictable windows.

Surveillance becomes most dangerous when it is blended with coercion. In controlling relationships, tracking is framed as love. As care. As protection. "Share your location so I know you're safe." "Let me see your phone so I can trust you." "If you have nothing to hide, it shouldn't matter." These statements sound reasonable until you recognize what they do. They collapse boundaries. They create ongoing entitlement. They turn privacy into guilt and access into a right.

Women are also tracked through the people around them. A child's tablet becomes a surveillance tool. A partner's shared account becomes a monitoring path. A workplace device becomes a sensor inside a woman's home. A friend tagging photos reveals location. A family member posting routines reveals patterns. Even when a woman is careful, she can be made visible by someone else's defaults. Surveillance rarely stays contained to the person who "accepted."

The systems that normalize tracking also normalize retention. They store history because it is useful. Useful for advertising. Useful for enforcement. Useful for "improving services." Useful for profiling. And once history exists, it becomes permanent leverage. A woman's past becomes searchable. Her movements become reconstructable. Her relationships become chartable. Her behavior becomes measurable in a way that can be interpreted without her.

The most frustrating part is that surveillance is treated as a security solution, even though it creates new risks. Monitoring does not guarantee safety. It guarantees records. A home camera does not prevent harm. It creates footage. A tracking app does not create trust. It creates permission. A platform log does not protect a woman's future. It creates a file that can be used later. Surveillance is sold as protection, but it often functions as control.

Women are expected to adjust. To accept that being tracked is normal. To accept that privacy is outdated. To accept that transparency is the price of participation. But women know in their bodies that constant observation changes behavior. It changes speech. It changes movement. It changes what feels safe to say. It changes what feels safe to search. It changes what feels safe to plan. Surveillance does not just record women. It shapes them.

This is the constraint: tracking became normal by being embedded into convenience, administration, and social expectation. Consent became routine and meaningless. Surveillance became invisible and constant. Women are tracked through devices, environments, institutions, and relationships. And the more the record grows, the more leverage exists over a woman.

The Shift

The shift begins when a woman stops imagining
surveillance as a rare event and starts seeing it as
infrastructure. Not a scandal. Not a special situation.
Infrastructure. Phones, apps, workplaces, cars, schools,
and homes are built to collect and retain information by
default. Once a woman sees that clearly, she stops
waiting for a moment where surveillance "goes too far."
It already went far. It just did it slowly, politely, and
under the cover of convenience.

Women are often told the answer is to calm down. That
tracking is just how modern life works. That privacy is
outdated. But women know what constant observation
does. It changes behavior. It narrows speech. It raises
vigilance. It creates self-censorship. It makes a woman
feel like she must perform normalcy to avoid trouble.
The shift is treating that feeling as evidence, not drama.
If a system changes how you live, it has power over you.

The first practical shift is rejecting the false choice
between total exposure and total isolation. Most women
cannot opt out. They still need to bank, work, schedule,
communicate, and move through public systems. The
goal is not disappearance. The goal is reduction. Less
collection. Less retention. Less sharing. Less automatic
visibility. A woman does not need perfect privacy to
regain control. She needs fewer open doors.

A woman starts by noticing what surveillance actually
looks like in daily life. It looks like location services
running all the time. It looks like "always allow"

microphone access. It looks like apps that want Bluetooth for "nearby devices." It looks like social platforms that default to public. It looks like cars and smart TVs collecting usage data. It looks like workplace tools logging activity. It looks like school portals tracking communication. Most of this is not hidden. It is simply buried in settings women are never encouraged to review.

This is where the shift becomes direct: women stop auto-accepting. A permission prompt is not a request from a neutral helper. It is a request from a system that benefits from more access. Women can pause. They can deny. They can set "only while using." They can turn off background location. They can restrict photo access. They can refuse contact syncing. These are not dramatic moves. They are boundary moves. They remove unnecessary data flow.

Women also reclaim power by treating tracking features as potential coercion tools, not safety tools. Location sharing can help in emergencies. It can also become a leash. Read receipts can support communication. They can also become demand and surveillance. "Last active" can feel casual. It can also become interrogation. The shift is understanding that a feature is not good or bad in the abstract. It is good or bad based on who can use it against you.

This matters because women are often pressured to "prove" trust through visibility. Share your location. Show your messages. Keep your phone unlocked. Let me have your passwords. Let me monitor you "for safety." These demands are normalized in relationships and families. The shift is recognizing that trust does not require access to your entire life. Healthy relationships do not require permanent monitoring. A woman can refuse surveillance without being dishonest.

Women also need to stop assuming that surveillance is only personal. It is institutional. Workplaces track devices. Apps track behavior. Portals store histories. Services retain logs. And when stress rises, those logs become enforcement tools. A system does not need to harm you intentionally to harm you. It only needs to store the record and allow others to query it later. Women who understand this stop treating records as neutral and start treating them as future leverage.

Another shift is realizing that some of the most dangerous tracking happens through other people. Children's devices. Shared accounts. Family photo sharing. Friends tagging locations. A woman can tighten her own settings and still be exposed through someone else's defaults. The shift is learning that privacy is not only individual. It is relational. It requires conversations about what gets posted, what gets shared, what gets tagged, and what gets saved.

The shift also includes recognizing that the enemy is not only "the watcher." It is retention. If tracking existed only for the moment, it would be less dangerous. The danger is that tracking creates history. History becomes narrative. Narrative becomes proof in systems that do not understand context. A woman who reduces retention reduces future risk. She creates fewer permanent trails that can be reconstructed when she needs to move quietly.

This is why women benefit from adopting a "privacy by design" mindset in daily life. Not perfection. Design. Before using a tool, a woman can ask: does this require constant collection to function? Does it store history automatically? Does it share with third parties? Does it expose location, identity, or relationships? If the answers are yes, she can choose a different tool, limit permissions, or keep the tool out of the center of her life. The shift is making those decisions before pressure arrives, not during it.

Women also regain autonomy by creating redundancy. Surveillance systems become most dangerous when they are tied to single points of failure. One phone number. One cloud account. One device. One platform. One workplace login. When those control everything, surveillance and control merge. Redundancy breaks that. A woman with backup access can leave an environment faster. She can recover faster. She can move without begging a system to restore her life.

The shift is not about living in fear. It is about refusing to be shaped by observation. Women have always adapted to environments that demand vigilance. Digital systems simply made that vigilance constant. Women can reduce the constant pressure by limiting what is collected, limiting what is visible, and limiting what becomes permanent. They do not need to win against surveillance. They need to make it less total.

This chapter explains how tracking became normal through convenience, administration, and social expectation. Surveillance is embedded into devices and platforms, then defended as safety, then normalized until women stop noticing the cost. The shift is learning to see tracking as infrastructure, then building boundaries that reduce collection and reduce retention. A woman does not have to disappear to reclaim control. She only has to stop donating continuous visibility.

The point is not to live untraceable. The point is to live ungoverned by default observation. When a woman reduces tracking, she regains freedom of movement. When she reduces retention, she reduces future leverage. When she refuses coerced visibility, she protects her boundaries. This is what digital self-defense looks like under surveillance as the default: not paranoia, but deliberate limits that keep her life hers.

~4
Risk Is Not Distributed Equally

The Constraint

People often talk about digital risk like it is universal. A password leak is bad for everyone. A hacked account is annoying for everyone. A doxxing incident is dangerous for everyone. The language is neutral, so the problem sounds equal. But risk is not distributed equally. The same exposure produces different consequences depending on who you are, how you are treated, and what power others feel entitled to take from you. Women do not experience digital risk as inconvenience. Women experience it as escalation.

Women live closer to consequence because women are more likely to be targeted. More likely to be stalked. More likely to be harassed. More likely to be punished socially for refusing attention. Digital exposure makes those threats cheaper and faster. It turns a stranger's curiosity into access. It turns harassment into persistence. It turns suspicion into a search bar. A woman does not need to be public-facing to be vulnerable. She only needs to be findable.

The first constraint is physical safety. For many women, digital exposure is not just embarrassing. It is dangerous. A location leak can become a follow. A photo background can become an address. A routine can become a schedule. An online argument can become offline attention. Women are taught to think of

43

safety as behavior management. Don't post too much. Don't argue. Don't respond. But the real issue is that digital systems create visibility that other people can exploit.

The second constraint is social punishment. Women are judged more harshly for the same digital behavior. A woman's tone becomes the story. Her boundaries become "drama." Her refusal becomes "attitude." Her privacy becomes "suspicious." She can lose credibility for protecting herself, because women are expected to be accessible. Digital life reinforces that expectation by making constant reachability feel normal.

Women are also more likely to be carrying other people's vulnerability. Children. Partners. Parents. Clients. Patients. Students. A woman's exposure often reveals information about others who did not consent. A school logo in a photo. A child's name in a message thread. A calendar appointment reminder. A medical portal notification. A location pinned to a favorite park. Women are not only protecting themselves. They are protecting a network.

This is why women's digital footprints are more valuable to exploit. A woman's device is often a control center. It contains the schedules, relationships, and logistics that keep life running. It contains not just what she wants, but what others depend on. That makes compromise higher impact. When a woman loses access, household stability shakes. When a woman is targeted, multiple people get pulled into the blast radius.

Women also face higher consequences because they are more likely to be trapped inside systems that require compliance. Schools require portals. Workplaces require authentication. Healthcare requires apps. Banking requires identity checks. If a woman is locked out, she cannot simply stop using the system. She still has to function. That need creates leverage for institutions and attackers alike. It creates pressure to restore access quickly, even if restoring access requires surrendering more data.

Digital risk also becomes reproductive and medical risk. Women's bodies are politicized. Women's healthcare is monitored. Women's choices are judged. When women's health data is collected and shared, it becomes more than a private detail. It becomes a vulnerability. It becomes a record that can be used against her socially, economically, or legally depending on the climate. Even when a woman trusts her doctor, she may be forced into digital portals and apps that expand her exposure beyond the exam room.

Women's work lives are also shaped differently by exposure. A man can be abrasive online and be called confident. A woman can be firm online and be called unstable. Women are evaluated through likability and compliance. That means online traces carry more weight. A screenshot can become "evidence." A misunderstood post can become a character assessment. Women face higher professional penalty for the same digital mess.

The constraint deepens because women are targeted by systems and by individuals at the same time. The platform may not protect her. The institution may not believe her. The social environment may blame her. She is told to document abuse, but the documentation becomes more exposure. She is told to report harassment, but the reporting process may require her to repeat details and provide proof that becomes permanent. Women are asked to perform credibility under pressure, and the system treats that performance as part of the cost of participation.

Women are also more likely to be punished when they try to leave. Leaving a relationship. Leaving a community. Leaving a job. Leaving a platform. Digital systems remember. Old accounts persist. Old messages persist. Old photos remain. Old contacts remain. The ability to disconnect cleanly is a luxury many women do not have. A woman can delete an account and still be searchable through cached data, shared screenshots, and other people's archives.

A woman's privacy is also contested inside her own life. She may be pressured to share passwords. Pressured to keep location on. Pressured to prove transparency. Pressured to surrender a device. These demands are often framed as love, safety, or trust. But they function as entitlement. They normalize constant access to a woman's life. When this happens, digital exposure is not accidental. It is enforced.

This is why the phrase "just log off" is meaningless advice for women. Logging off does not erase the record. Logging off does not end institutional monitoring. Logging off does not stop harassment if the harasser already has information. Logging off does not remove the power imbalance that makes women more vulnerable to escalation. Women cannot solve structural risk with personal retreat.

This is the constraint: women live with higher stakes. Digital exposure hits harder because women are more targeted, more judged, more burdened with relational responsibility, and more restricted by systems that require compliance. Risk is not distributed equally, and pretending it is equal makes women carry the consequences alone.

The Shift

The shift begins when women stop demanding that digital life be fair and start building strategies based on what is true. Fairness is not the operating principle of modern systems. Scale is. Convenience is. Compliance is. And when those priorities collide with women's safety, women are usually the ones asked to absorb the cost. Not because women are weak, but because women are expected to be flexible. Expected to be polite. Expected to adapt. The shift is refusing to treat that expectation as natural law.

The first shift is naming the reality without apology: women face higher consequences. Not hypothetical ones. Real ones. Stalking, harassment, coercion, workplace penalty, family pressure, social punishment, legal risk, and physical danger. When women minimize that truth, they end up building protection for a fantasy world. A fantasy where reporting works. A fantasy where platforms respond quickly. A fantasy where institutions understand context. A fantasy where "just block them" ends the problem. Digital self-defense begins when women build for the world they actually inhabit.

That changes the goal. The goal is not "feeling safe online." The goal is maintaining autonomy when someone tries to exploit exposure. Autonomy means a woman can keep moving even when pressure rises. She can keep access to her money. She can keep communication pathways open. She can keep control of her identity. She can prevent a single breach or single mistake from collapsing her life. Digital protection is not a product. It is an architecture of options.

Another shift is recognizing that women's risk is rarely isolated. Women are rarely protecting only themselves. They are protecting children, parents, partners, clients, patients, students, and communities that depend on them. A woman's phone is often a household control center. That means a woman's security decisions are not just personal preferences. They are protective decisions. When women treat them that way, they stop second-guessing the seriousness of boundaries.

Women also stop accepting shame as a management tool. Shame is how systems and people push women into compliance. "If you have nothing to hide, it shouldn't matter." "Why are you being difficult?" "You're overreacting." "You're paranoid." "You're dramatic." These lines exist to make women surrender. The shift is recognizing shame as a tactic. A woman can refuse exposure without debate. She can refuse to explain why she wants less visibility. She can treat privacy as normal, because it is normal.

This matters because women are often trained to trade privacy for social comfort. Be reachable. Be responsive. Share details. Confirm plans publicly. Keep location on so people don't worry. Keep read receipts on so people don't get mad. Keep social presence so you don't look suspicious. But convenience for others becomes surveillance of you. The shift is refusing the belief that women owe access in order to be considered trustworthy.

Women build better protection when they think in layers. There is personal risk from individuals. There is institutional risk from systems. There is economic risk from dependency. There is social risk from exposure. Women are targeted at all of these layers. So the solution is not one "secure app" or one settings change. The solution is reducing fragility at the points where consequences concentrate.

That starts with access. A woman's life collapses fastest when she loses access to core accounts. Email. Phone number. Banking. Device unlock. Cloud storage. Password recovery. When women prioritize these, everything else becomes less terrifying. If harassment escalates, she can still communicate. If a platform fails, she can still log in elsewhere. If someone tries to lock her out, she can still recover. The shift is treating access points like critical infrastructure.

The next layer is visibility. Women reduce their footprint where exposure creates real-world risk. They do not need to erase their presence. They need to remove the easy pathways. Location sharing defaults. Public profiles with full names. Tagging and automatic face recognition. Routine posts that reveal patterns. Profiles linked across platforms. Children's information in public feeds. None of these are moral issues. They are safety issues. The shift is choosing what is private by design instead of private by hope.

Another layer is relational boundaries. Women are pressured to share devices, passwords, and location access inside families and relationships. That is often framed as trust. But trust does not require permanent monitoring. The shift is recognizing the difference between coordination and surveillance. Coordination is voluntary and limited. Surveillance is ongoing entitlement. A woman can share what is necessary for safety and refuse what is demanded for control.

Women also shift by treating platforms as unstable environments, not as community anchors. Social platforms change rules. Moderation fails. Accounts get suspended. Harassers evade blocks. Reporting systems do not protect women reliably. The shift is building a life that does not depend on one platform for income, identity, community, or safety. A woman does not need to abandon social spaces. She needs redundancy, so she can leave without losing her world.

This is where women stop trusting the idea of "proof" as protection. Women are told to document abuse. Save screenshots. Keep logs. Report it. Sometimes that is necessary. But the record can also become exposure. It can spread. It can be dismissed. It can be turned back on her. The shift is understanding that documentation is not the same as safety. It is one tool in a system that often fails women. Women document when it helps, but they do not confuse documentation with protection.

Women also stop taking on the job of convincing institutions to care. In theory, institutions respond. In practice, women are often forced into credibility performances: explaining, proving, repeating, staying calm, providing evidence, meeting deadlines, filling out forms, and enduring delay. Meanwhile, the harassment continues. The shift is building protection that does not require institutional permission. Women aim to reduce what can be exploited before they are forced into the system's slow response loop.

The most important shift is moving from reaction to posture. Reaction is what systems force women into. An incident happens, and she scrambles. She spends hours fixing settings, recovering access, explaining herself, managing fallout. Posture is building a default environment that is harder to exploit. It is reducing exposure before harassment starts. It is building redundancy before lockout happens. It is separating accounts before compromise spreads. Posture is calm preparation so women do not have to negotiate with panic later.

This chapter is about recognizing that women are not "overreacting" to digital risk. Women are responding to a world where risk concentrates on those who are most targeted and most constrained by consequence. The constraint is not only technological. It is social and structural. Women face higher penalties for exposure, higher vulnerability to escalation, and higher pressure to stay accessible. The shift is building layered protection that prioritizes access, reduces visibility, strengthens boundaries, and creates redundancy.

The point is not to live in fear. The point is to live with accurate assumptions. A woman who accepts that risk is not equal can stop copying generic advice never designed for her. She can build protection that matches her reality. She can keep her life running while reducing the leverage others can take. That is what autonomy looks like when the world is not fair: durable control when pressure arrives.

~5
Control Versus Convenience

The Constraint

Women are told convenience is progress. Faster logins.
Easier payments. Automatic syncing. Smart suggestions.
A single tap to approve, a single tap to store, a single
tap to share. The pitch is that modern life is finally
getting simpler. But convenience is rarely neutral.
Convenience is a trade. And the system sets the terms of
that trade so it does not feel like a decision.

Convenience is how control enters without resistance. If
a woman is offered something that saves time, she is
expected to accept. If she hesitates, she is treated as
paranoid or difficult. If she refuses, she is punished with
friction. Longer steps. More verification. More errors.
More delays. The system does not argue with her. It
simply makes refusal exhausting.

The first constraint is that women are rarely given real
choice. Most "choices" are designed as forced options.
Agree or lose access. Share your contacts or miss key
features. Turn on notifications or fall behind. Use the
app or you cannot complete the task. These are not
choices. They are gates. And once gates are normalized,
consent becomes a ritual instead of a meaningful
decision.

Women are also pressured to accept convenience
because they are managing time for everyone. Women

are coordinating children, work, family obligations, and household logistics. When a system offers speed, it looks like relief. But relief becomes dependency quickly. A woman accepts a shortcut today, and tomorrow that shortcut becomes the only path. Then the system owns the pathway. This is how convenience becomes control. It collapses redundancy. It pulls everything into one login, one platform, one cloud, one device, one identity anchor. The marketing language is "streamline." The reality is centralization. Centralization makes life easy when the system works. It makes life fragile when it doesn't.

Most women discover fragility when they hit a wall. A device breaks and half of life is locked behind recovery steps. A phone number changes and account access collapses. A platform suspends an account and work tools disappear. A payment system flags a transaction and money freezes. These failures feel personal because they interrupt a woman's life directly. But the failure was structural. The system was designed to be convenient at the cost of resilience.

Convenience also reframes exposure as comfort. Location sharing becomes "safety." Cloud syncing becomes "peace of mind." Smart home devices become "security." Password managers become "ease." Biometrics become "simplicity." Some of these tools can help women. The issue is how the framing hides the tradeoffs. A woman is told what she gains. She is not told what she loses.

What women lose is control over the record. Convenience creates permanent history. It saves everything because saving is part of the product. The system stores messages, photos, searches, locations, purchases, and contacts because storage increases value. And once stored, that record becomes portable. It can be merged. It can be sold. It can be queried. It can be subpoenaed. It can be used in decisions the woman never agreed to participate in.

Convenience also changes the boundaries of privacy. A woman is pushed to keep devices always listening, always syncing, always tracking, always connected. "Always on" becomes the new baseline. And when always on becomes normal, the idea of privacy starts to feel like a disruption. Women are made to feel like privacy is selfish, or outdated, or suspicious. That is how control works: it makes refusal feel abnormal.

Women are also targeted through emotional convenience. Platforms offer quick validation. Quick connection. Quick responses. Quick entertainment. The system teaches women that attention is relief. It trains them to keep checking. Keep responding. Keep staying visible. Over time, that becomes a form of control too. A woman's attention becomes a resource that can be harvested and redirected. Convenience becomes compulsion.

The deeper constraint is that convenience removes skill and replaces it with dependency. When everything is automated, women lose familiarity with how systems work. They don't know how backups operate. They don't know how recovery works. They don't know where data lives. They don't know what is connected to what. Automation is sold as freedom, but it often creates helplessness. The woman cannot intervene because she does not control the machinery.

This is why convenience is so dangerous in authoritarian climates. Control does not need to arrive as violence. It can arrive as infrastructure. When identity, communication, and payment systems are centralized and monitored, enforcement becomes easy. Access can be revoked. Accounts can be flagged. Services can be denied. A woman can be contained through normal procedures. The system calls it policy. The woman experiences it as confinement.

Even in non-authoritarian environments, convenience still concentrates power. It makes companies and institutions the gatekeepers of recovery. A woman is told her account is protected, but protection is conditional. The system will "help" her only if she can satisfy verification steps. And those steps often assume a stable life. Stable phone numbers. Stable addresses. Stable devices. Stable work histories. Women living through instability pay more.

Women are told that convenience is optional. But in practice it becomes mandatory. Schools require portals. Employers require authentication apps. Healthcare requires online accounts. Banking requires verification. Travel requires digital records. A woman can resist some systems, but she cannot resist all of them without losing access to basic services. This is how choice is removed: not through force, but through routing.

Control is not just about surveillance. It is about shaping behavior. Convenience creates the smooth path, and then the smooth path becomes the only path. A woman is guided into compliance because compliance is what works. Over time, she stops questioning it. She simply adapts. She accepts friction as normal. She accepts exposure as the cost of functioning. She accepts that her life is stored in systems she cannot fully audit.

This is why digital self-defense cannot be framed as "just use better tools." Better tools help, but the deeper problem is the structure of the trade. Women are being trained to exchange autonomy for speed, visibility for access, and privacy for convenience. And those exchanges are often irreversible. Once data is collected, it cannot be uncollected. Once dependency is built, leaving becomes expensive.

Most women are not told they are making a trade. They are told they are getting a benefit. Faster. Easier. Seamless. One click instead of ten. Convenience is framed as progress, so refusing it feels irrational. But convenience is not free. It is a design strategy that often hides control behind ease, and the tradeoff is usually paid in privacy, dependency, and reduced choice.

The constraint is that the system presents convenience as the only reasonable path. If women accept, life moves smoothly. If women hesitate, friction appears. Extra steps. More prompts. More warnings. More delays. Over time, women learn that saying yes is the fastest way to keep life running. That is how control enters. Not through force, but through routing.

The Shift

The shift begins when women stop treating convenience as a gift and start treating it as a contract. A contract always has terms. It offers something now in exchange for something later. Speed in exchange for data. Ease in exchange for dependency. Access in exchange for visibility. The system rarely says this directly. It packages the trade as progress. But women feel the cost anyway. They feel it when they try to step back and realize there is no way back.

Most women have been trained to accept convenience automatically because modern life is built to punish hesitation. Time is scarce. Responsibility is constant. Women are coordinating children, work, family, and household stability. When a screen offers "one click," it looks like relief. But relief becomes default. Default becomes dependence. And dependence becomes control when pressure rises.

This is why women need a different posture. Not rejection. Not paranoia. Posture. Women can keep convenience where it helps and refuse it where it creates leverage. The shift is recognizing that real choice is not what the screen offers. Real choice is what still works when you say no. Women regain real choice by learning to recognize a gate disguised as a feature. A gate is a system that makes the "optional" thing required. It says you can skip, but the service breaks if you do. It says you can decline, but it punishes you until you comply. It says you can opt out, but it routes you back in. Women can name these patterns and stop interpreting them as personal inconvenience. The friction is the point.

This changes how women use tools. Instead of asking "Is this easy?" women ask, "What does this make dependent?" Instead of asking "Is this popular?" women ask, "What happens if this fails?" Instead of asking "Will this save me time?" women ask, "What will this cost me later?" These questions do not require obsession. They require honesty about consequences.

One of the cleanest moves women can make is restoring redundancy. Convenience collapses redundancy by design. One login for everything. One device for everything. One cloud for everything. One password reset method. One platform that holds identity, communication, photos, files, payments, and recovery. The system calls it streamlined. But streamlined is fragile. The shift is insisting on at least one backup route for anything that would collapse your life if it disappeared.

This is not about building a bunker. It is about removing single points of failure. A woman who has one pathway is governable. A woman who has two pathways has options. Options are the foundation of autonomy. Options are what make pressure survivable. Women also regain control by reducing automatic sharing. Convenience loves "always-on." Always-on location. Always-on syncing. Always-on cloud backups. Always-on contact sharing. Always-on microphones. Always-on notifications. It feels smooth until the day it becomes exposure. The shift is moving from always-on to intentional. A woman can choose "only while using."

She can choose manual upload instead of full library access. She can choose to stop syncing what does not need to be synced. She can decide what belongs in the permanent record and what does not.

Women can also stop confusing convenience with safety. Many systems sell monitoring as protection. Location sharing becomes "I'm just making sure you're okay." Read receipts become "I just want to communicate." Device access becomes "I just need reassurance." But convenience can be a mask for entitlement. The shift is recognizing the difference between voluntary coordination and forced transparency. Safety does not require permanent observation.

This matters because women are often pressured to perform trust by giving up privacy. Share passwords. Share your phone. Keep location on. Provide proof. Stay reachable. Women are told that boundaries create conflict. But boundaries do not create conflict. Entitlement creates conflict. The shift is refusing the belief that women owe access to be believed. A woman can say no and treat that no as normal. Women also regain choice by simplifying their own rules. Convenience wins when women are tired and overloaded. So the strongest defense is not complicated. It is repeatable. If a woman has to make a new decision every time, she will eventually default to the fast option. But if she has a few consistent rules, she stays stable under pressure.

A woman can decide: one place for money, and it gets the highest protection. One place for private communication, and it does not get casual sharing. One place for cloud storage, and it does not become a dumping ground for everything. One place for photos, and it does not automatically sync to every app. These are not technical rules. They are boundaries that reduce chaos.

Another shift is accepting that convenience often hides permanence. One-click doesn't just complete a task. It stores history. It saves identifiers. It creates records that can be merged later. Autocomplete does not just help you type. It trains the system on your habits. Smart suggestions do not just reduce effort. They reduce your independent decision-making. Convenience is often a way to move your life into the system's memory.

Once women recognize permanence, they can start deleting the idea that "it's no big deal." It becomes a big deal when the record is queried later. When someone uses it to profile you. When an institution treats it as proof. When a relationship turns controlling. When the political climate shifts. The shift is thinking forward without spiraling. Not imagining worst cases. Simply recognizing that history becomes leverage.

Women also need to stop waiting for a system to rescue them when convenience breaks. Platforms and services do not exist to protect women. They exist to retain users and reduce liability. If a woman's account is locked, the system's job is to follow procedure. If she is harassed, the system's job is to reduce public scandal, not necessarily to reduce her harm. The shift is building autonomy that does not depend on corporate goodwill.

This is why women benefit from designing their digital life for exit. If a platform becomes unsafe, she can leave without losing everything. If a device is compromised, she can replace it without losing access. If an account is suspended, she can keep working. Exit is not pessimism. Exit is power. When women have exit routes, convenience loses its ability to trap them.

This chapter is about restoring real choice in an environment designed to narrow it. The constraint is that convenience is framed as progress while functioning as control: it centralizes access, reduces redundancy, increases retention, and trains compliance through friction. The shift is treating convenience as a contract and reading the terms. Women keep convenience where it serves them and refuse it where it creates leverage. They rebuild redundancy, reduce always-on sharing, and design their digital lives for recovery and exit.

The point is not to become rigid. It is to become unmanageable by default settings. When women slow down just enough to notice the trade, they get choice back. When women keep a backup route, they reduce fragility. When women stop auto-accepting "easy," they stop donating control. Convenience is useful. But control is non-negotiable.

~6
Digital Harm Is Cumulative

The Constraint

Many women are trained to think digital harm is a single event. A hack. A leak. A stalker. A viral post. A humiliating screenshot. Something obvious, dramatic, and easy to identify. But most harm does not arrive as one explosion. It arrives as accumulation. A little exposure here. A little compromise there. A few weak defaults that never get fixed. Over time, those small openings form a system of vulnerability.

The first constraint is that small exposures feel harmless in isolation. One app gets location access. One platform gets contact syncing. One website stores a password. One service keeps payment info on file. One device stays logged in because it is easier. None of these choices feels like danger. They feel like normal participation. But harm is not always the immediate result of a single choice. Harm is often the outcome of many choices becoming a pattern.

Digital systems are designed to encourage this accumulation. They ask for access gradually. They normalize small permissions. They reward convenience. They default to retention. They store data because storage creates value. And once stored, the data does not stay inside one app. It becomes portable. It can be copied, shared, sold, merged, and retained long past the moment a woman thought it mattered.

Women are also taught to treat risk as binary. Either you are safe or you are unsafe. Either you got hacked or you didn't. Either someone is stalking you or they aren't. This framing hides the reality that risk is often incremental. A woman can be partially exposed in ways that do not trigger crisis but still reduce her freedom over time. Harm can be present even when everything still "works."

This is why cumulative harm is so hard to see. A woman may not feel threatened today. She may feel fine. She may have nothing obvious happening. But the system is still building a record. The record grows quietly in the background: location history, message history, contact graphs, search behavior, photos, purchases, social interactions, device identifiers. The woman experiences the cost later, when she tries to change course and realizes how much of her life has been stored.

A small exposure also becomes larger through merging. One account leak reveals an email. That email is linked to other accounts. A reused password opens another door. A phone number connects identities across platforms. A photo reveals a workplace. A workplace reveals a city. A city reveals a neighborhood. A neighborhood reveals a school. None of this requires a sophisticated attacker. It requires time and access to common tools. That is the real danger of accumulation: it lowers the barrier.

Women face higher cumulative risk because women are expected to remain visible and reachable. A woman is expected to respond quickly. Expected to keep accounts active. Expected to be findable. Expected to coordinate through apps. Expected to maintain digital presence for work and family. That constant participation creates constant data. And constant data creates a constant trail.

Cumulative harm also shows up as fatigue. Women are asked to update passwords. Approve logins. Confirm identities. Manage notifications. Filter spam. Block strangers. Report abuse. Fix broken settings. The system offloads the labor onto women while continuing to collect. Each small task feels manageable. The total load becomes heavy. Over time, women get tired and stop tightening the gaps. That is when cumulative exposure becomes cumulative vulnerability.

The harm is not only exposure. It is loss of agency. When a woman's life is stored across systems, she loses control over her own narrative. Other people can reconstruct her history. Institutions can query her patterns. Platforms can infer her behavior. A woman can be judged by records that are incomplete, outdated, or wrong. But the record still has authority because it is digital. Digital history is treated as proof even when it lacks context.

Cumulative harm also changes what women feel safe doing. A woman learns to speak differently because screenshots last. She learns to move differently because location can be tracked. She learns to plan differently because messages can be read. She learns to keep things off the record because the record does not forget. That shift in behavior is harm. It is harm even when nothing overt happens, because it narrows a woman's freedom.

Women are also harmed cumulatively by small social exposures. A friend tags a location. A family member posts a photo. A group chat shares a detail. A coworker adds a workplace reference. A child uses a real name in a public game. These moments seem minor. Together, they create a fuller picture than a woman ever intended to share. Women can try to control their own accounts and still become visible through other people's defaults.

Another constraint is that platforms profit from cumulative exposure. They are built for retention. They reward habit. They reward oversharing. They reward engagement. They do not reward women for leaving. They do not reward women for reducing data. They do not reward women for minimizing. A woman is swimming upstream when she tries to reduce her footprint, because the system is designed to pull her back into accumulation.

Cumulative harm is also how digital coercion becomes normal. A woman may accept one boundary violation. One request to share a password. One request to keep location on. One request to prove something with screenshots. One demand to be reachable at all times. These demands can start small. Over time, they become entitlement. The woman is trained to believe she must stay transparent to keep peace. That is cumulative harm. It is control built slowly enough to feel normal.

The hardest part is that this kind of control rarely announces itself as control. It presents as concern. As love. As partnership. As responsibility. "I worry about you." "I just want to know you're okay." "I'm only asking because I care." The framing makes refusal feel cruel, even when the request is about access, not safety. Women are pushed to trade boundaries for harmony, and over time that trade becomes expected.

Once entitlement is established, the system reinforces it. A woman's phone becomes a permission structure for someone else's anxiety. Her messages become evidence. Her timestamps become arguments. Her silence becomes suspicion. The digital record turns normal life into something that can be audited. And because the record exists, the demand grows: prove it again. Explain it again. Share more. Stay reachable. Stay visible.

This is the constraint: digital harm is rarely one event. It is the accumulation of small exposures, small compromises, small defaults, and small losses of boundary. The record grows. The systems merge it. The social environment reinforces it. Women carry the labor of managing it, and women carry the consequences when the accumulated trail becomes leverage.

The Shift

The shift begins when women stop looking for the one dramatic threat and start paying attention to slow damage. Cumulative harm is not loud. It does not arrive with one moment that makes everything obvious. It arrives as small exposures that feel normal, small compromises that feel manageable, and small defaults that feel like "just how things work." A woman can feel fine while harm is building. The shift is learning to treat that as the real problem.

Most women are trained to manage risk as a crisis response. Something happens, and then she scrambles. She changes passwords. She blocks accounts. She reports harassment. She deletes posts. She resets settings. That response is sometimes necessary. But cumulative harm is not solved through emergency moves. It is solved through posture. A posture is what a woman does before there is a crisis. It is how she builds a life that does not steadily leak power without noticing.

This is why the most important move is naming accumulation as the mechanism. Exposure does not have to be extreme to become dangerous. A little location history builds a routine. A few leaked identifiers link identities. A few photos reveal patterns. A few apps with extra permissions create multiple points of collection. A few years of stored messages create a searchable archive. The shift is seeing that harm comes from volume and longevity, not just intensity.

Women regain leverage by reducing what can accumulate. Not everything. Just the things that matter most. Location history. Photos. Messages. Contact syncing. Password recovery methods. Public profiles tied to real identity. If those are tightened, the rest becomes less risky. Women do not need to fix the entire internet. They need to stop donating permanent, mergeable records to systems that cannot be trusted to hold them safely.

The shift also includes refusing to treat convenience as an excuse for permanent storage. A woman can choose to keep some things off the record on purpose. Not because she is hiding wrongdoing, but because she is protecting future flexibility. She does not need every photo auto-synced forever. She does not need every conversation stored indefinitely. She does not need every location ping archived. The ability to forget is a form of freedom. Cumulative harm is what happens when forgetting becomes impossible.

Women also benefit from making separation normal. When everything lives in one place, accumulation becomes collapse. One cloud account holds messages, photos, documents, backups, contacts, and location history. One device holds everything. One identity anchor links platforms. That is efficient. It is also fragile. Separation reduces blast radius. It prevents one compromise from becoming total exposure. This matters because cumulative harm is often invisible until a woman needs to move. A move can be literal or social. She may need to leave a relationship. Leave a job. Leave a community. Change routines. Protect a child. Reduce contact. In those moments, the past becomes leverage if the record exists everywhere. The shift is building a life where leaving is possible because exposure is limited and dependencies are not total.

Women also regain leverage by adopting a maintenance mindset. Digital harm builds like mold. It spreads in neglected corners. It grows where no one looks. A woman does not need to constantly panic-check settings. But she does need a baseline of upkeep. Remove what is no longer needed. Tighten what drifted open. Turn off what does not need to run. Reduce the number of systems that have access to her life by default. Maintenance is how cumulative harm is interrupted.

Another shift is learning that boundaries need to be consistent, not perfect. Women often fail at digital protection because they try to do it all at once. They attempt a total overhaul, get overwhelmed, and return to old defaults. The more sustainable move is incremental tightening. One permission removed. One setting changed. One account separated. One habit adjusted. Over time, small defensive choices compound the way exposure does. The difference is that this compounding works for the woman, not against her.

Women also have to treat relational coercion as cumulative risk, not interpersonal drama. Control rarely begins as a demand. It begins as a question that is hard to refuse. "Can you share your password?" "Can you leave your location on?" "Can you show me the messages?" "Can you prove where you were?" The request is framed as concern. As trust. As safety. The shift is recognizing that repeated requests for access are not reassurance. They are entitlement training. Women can choose to refuse early, when the cost is lower. The longer coercion goes unchallenged, the more normal it becomes. Then refusal looks like betrayal, even when it is simply a boundary. The shift is understanding that privacy is not selfish. Privacy is necessary. A woman can have a relationship without surrendering her entire digital life for inspection.

Women also stop assuming that documentation is the same as safety. Women are often told to screenshot everything, save everything, keep records, keep proof. Sometimes that is necessary. But documentation is also accumulation. It creates more files, more archives, more material that can be leaked or misused later. The shift is choosing documentation intentionally. A woman records what is needed, and then stores it in a way that is controlled. She does not turn her entire life into evidence just in case someone demands it.

The shift includes reclaiming the right to delete. Deletion is often framed as suspicious. But deletion is hygiene. It is how women reduce the amount of past life that can be pulled forward and used against them. A woman does not owe permanent records to anyone. She can reduce what exists. She can reduce what is visible. She can reduce what is searchable. The more she reduces, the less cumulative leverage exists.

Women also need to stop letting platforms define the end of a situation. A woman can delete an account and still be exposed through screenshots, caches, and other people's archives. That is why cumulative harm cannot be solved through platform settings alone. The shift is recognizing that the record lives in multiple places and that control requires multiple layers: personal boundaries, platform choices, permission choices, and social practices. Women are taught to treat convenience as the goal. The safer goal is stability. Stability is what remains when things break. A woman who reduces accumulation is not trying to live perfectly. She is trying to avoid being cornered by the past, and preserve the ability to move forward without dragging every prior version of herself behind her.

This chapter is about understanding that cumulative harm works because it is quiet, gradual, and normalized. Small exposures feel harmless until they merge into a complete picture. Small compromises feel manageable until they become entitlement. Small defaults feel irrelevant until they create permanent records that cannot be undone. The shift is interrupting accumulation through minimization, separation, maintenance, and consistent boundaries.

The point is not to eliminate risk. The point is to stop bleeding leverage. When women reduce what is collected, stored, and merged, they regain freedom. When they keep boundaries consistent, coercion stays small instead of becoming normal. When they design their digital lives for exit and recovery, the past loses its ability to trap them. Cumulative harm is slow. So is cumulative protection. And cumulative protection is how women take power back without having to disappear.

~7
Protection Without Withdrawal

The Constraint

Women are often taught that digital safety has only two modes. Either you stay online and accept the risk, or you disappear and lose access to modern life. Either you participate fully, or you protect yourself by withdrawing. That false choice keeps women trapped. It makes safety feel like isolation. It makes engagement feel like recklessness. And it makes women believe that if they want privacy, they must accept loneliness, inconvenience, and loss.

Women cannot realistically withdraw. They need technology for work, school, money, healthcare, childcare coordination, and communication. They need search tools. They need maps. They need portals. They need banking apps. They need group chats. They need email. They need access to information quickly because responsibility demands it. The world does not offer women the luxury of opting out without consequence.

At the same time, women cannot afford to pretend the risks are imaginary. A woman who stays fully visible becomes easy to target. Easy to track. Easy to pressure. Easy to harass. Easy to profile. And when women are harmed online, the consequences do not stay online. Exposure becomes escalation. A small leak becomes stalking. A screenshot becomes workplace damage. A location clue becomes a physical threat. Digital life can be weaponized fast.

This is the constraint: women are pushed into participation without protection. Platforms are designed to increase reach, increase visibility, and increase retention. They reward oversharing. They reward constant presence. They reward responsiveness. They do not reward boundaries. They do not reward minimization. They do not reward women for being careful. The system encourages behavior that increases exposure, then blames women when exposure is exploited.

Women are also pressured socially to remain accessible. A woman who is not reachable is treated as rude. A woman who does not respond quickly is treated as suspicious. A woman who does not share is treated as cold. Digital culture trained people to expect constant availability, and women absorb that expectation more than anyone. Women are expected to be the social glue. The planner. The responder. The emotional regulator. That social pressure turns engagement into obligation.

The problem becomes sharper because women do not only need technology for public life. They need it for private survival. Women use digital tools to find resources. Research legal options. Compare services. Ask for help. Find community. Learn skills. Build income. Plan moves. Coordinate safety. Those are not optional. But the same tools that support women also generate records that can be used against them.

Women are also taught that safety means being careful with obvious strangers. Don't respond to weird messages. Don't click suspicious links. Don't accept unknown friend requests. Those are real risks. But most harm comes from normal systems. Default settings. Data retention. Location leakage. Oversharing by friends. Account linking. Password recovery pathways. These risks are not addressed by "be careful." They are structural. Women can behave perfectly and still be exposed by the way the environment is built.

A woman can do everything "right" and still be vulnerable because the system is designed to collect first and explain later. A platform can change privacy defaults overnight. A phone update can re-enable tracking. A new feature can publish old information in a new format. A tool that was once private can become searchable. Women are blamed for not keeping up, even though the environment shifts faster than any individual can track.

Women are also exposed through other people's convenience. A friend tags a location without thinking. A relative posts a photo with a background detail. A coworker uploads a group picture with names attached. A child shares a real name inside a game. The woman did not consent, but she still becomes visible. Modern exposure is not only what a woman shares. It is what a network leaks around her.

The constraint becomes worse because most "safety" tools are built for platform stability, not women's safety. Reporting systems are slow. Blocking is incomplete. Harassers create new accounts. Platforms prioritize engagement metrics over containment. A woman is told to use the tools provided, then learns that the tools mainly reduce the platform's responsibility. Women end up doing the labor of protection inside systems that were not built to protect them.

This leaves women in a constant tension. They feel unsafe, but they still have to function. They want to speak, but they know speech can be saved. They want community, but community can become contact pressure. They want visibility for work, but visibility attracts unwanted attention. They want to help others, but helping can make them searchable. Women are trying to live normal lives inside systems that punish normal boundaries.

Platforms intensify this tension because they collapse audiences. A woman's friends, coworkers, relatives, clients, and strangers can all see the same content. One post can be interpreted by multiple groups with different expectations. That creates risk even without harassment. A woman's words can be taken out of context, screenshot, shared, and used as proof of a story she did not intend to tell. The platform calls this engagement. Women experience it as fragility.

The constraint also includes the reality that women are more likely to be targeted for refusing access. When women block someone, they may escalate. When women stop responding, they may intensify. When women set boundaries, they may be labeled hostile. Women are punished for leaving conversations that were never safe. The system offers blocking tools, but it does not offer true protection. It offers friction management, not safety.

Women are also pressured into using identity-linked systems. Real names. Verified profiles. Phone-number based logins. Photo-based identity checks. These are marketed as reducing abuse. Sometimes they do. But they also increase stakes for women. Identity linkage makes women easier to find across platforms. It makes their offline lives easier to connect to online traces. It reduces their ability to speak freely without consequences. Women are asked to trade anonymity for "safety" and then are blamed when the trade creates new vulnerabilities.

This is why the "just stay off social media" advice fails. It ignores work realities. It ignores family coordination. It ignores education, healthcare, and financial access. It ignores the fact that withdrawal can be its own harm. A woman who disappears loses community. Loses information. Loses opportunities. Loses voice. Women already live in a world where they are often made small. Withdrawal can reinforce that shrinking.

This is the constraint: women are being forced to choose between engagement and protection because systems are built to treat visibility as the default. The environment encourages women to be reachable, searchable, and legible. It increases exposure through design and then treats the fallout as an individual woman's problem. Women are not reckless. They are navigating a system that makes normal participation risky.

The Shift

The shift begins when a woman stops treating safety as disappearance and starts treating it as design. Withdrawal is not the only form of protection. It is simply the most obvious one. Women are often told that the internet is dangerous, so the answer is to stop being there. But women do not get to stop being there. Work, family coordination, healthcare, school systems, banking, and basic communication are digital now. The shift is learning how to stay engaged without turning engagement into exposure.

Most women already understand this in physical life. They go out, but they lock doors. They socialize, but they keep boundaries. They travel, but they keep backup plans. They do not solve risk by never leaving the house. They manage risk by controlling access. Digital life requires the same mindset. A woman can participate without surrendering full visibility. She can speak without storing everything forever. She can be reachable without being available to everyone.

The shift starts with rejecting the idea that a woman owes accessibility. Digital culture tries to turn "reachability" into a moral standard. If you do not respond quickly, you are rude. If you do not share details, you are cold. If you limit access, you are suspicious. Women absorb this pressure because women are trained to maintain social smoothness. But constant availability is not kindness. It is a vulnerability. A woman can be caring and still be private. She can be responsive and still be selective.

This is where women reclaim the power of deliberate friction. The system trains women to avoid friction at all costs. One click. One tap. Always-on settings. Automatic sharing. Seamless syncing. But friction is not always a problem. Sometimes friction is protection. A woman can choose slower methods where they reduce exposure. She can choose steps that force a pause before something becomes permanent. She can choose settings that limit what gets collected, stored, and shared. The shift is using friction on purpose instead of being punished by it.

Women also build protection by separating audiences. Platforms collapse audiences by default, which makes every post riskier than it needs to be. The shift is being intentional about where different parts of life exist. One space for work. One space for personal communication. One space for public writing. One space for private relationships. Separation is not hiding. It is sanity. It reduces the chance that strangers, coworkers, and family members can all interpret the same content through different expectations.

This becomes even more important when women are using online spaces for income or mission-driven work. Women need visibility to build an audience, sell products, share ideas, and create opportunity. The shift is understanding that visibility must be paired with containment. A woman can be public without being personally accessible. She can create value without giving strangers a direct route to her private life. She can publish without linking every identity thread into one searchable profile.

Women also stay engaged by treating identity as adjustable, not total. Many platforms push women toward full verification and full linkage. Real names. phone numbers. face-based identity checks. The framing is safety. But for women, full identity linkage increases stakes. It makes doxxing easier. It makes stalking easier. It makes offline consequences more likely. The shift is being selective about where full identity belongs and where it does not. A woman can choose environments where real identity is necessary, and keep other environments intentionally separated.

A woman also stays engaged by controlling discoverability. Discoverability is what makes participation risky. If you are searchable, you are targetable. If your profiles are linked, you are traceable. If your routines are visible, you are predictable. The shift is reducing the easy pathways. Not by disappearing, but by limiting what can be connected. Less public metadata. Less casual location sharing. Less cross-platform linking. Less permanence where permanence creates leverage.

Another part of the shift is shifting from reactive blocking to proactive boundaries. Blocking is useful. But blocking often comes after harm begins. Women need boundaries that reduce the chance of harm gaining traction. They can limit who can message them. Limit who can comment. Limit who can tag them. Limit what is public by default. Limit how much personal detail exists in the first place. This is not about fear. It is about removing the low-effort paths that attackers use.

Women also benefit from normalizing the idea of "safe engagement." Safe engagement means choosing the level of interaction that matches the environment. A woman does not need to treat every platform as equally safe. Some spaces reward chaos. Some spaces reward harassment. Some spaces are built with better containment. The shift is choosing where to speak, where to observe, and where to disengage without guilt. Engagement is not an obligation. It is a tool.

The shift also includes refusing to do safety labor alone. Women are often expected to manage all risk privately. They are told to "be careful" and "avoid drama." But women are not the cause of the threat. The threat is. The shift is building community practices that protect women without forcing women into isolation. Friends avoid tagging location. Family stops posting children's details. Workplaces respect boundaries. Communities treat privacy as normal. Women should not have to fight their own people to stay safe.

Women also regain power by designing for exit. Exit is not negativity. Exit is autonomy. Platforms change policies. Communities shift. Moderation fails. Accounts get suspended. Harassment escalates. Women who design for exit can leave without collapsing. They keep copies of important work. They keep backup contact routes. They keep financial access separate from platform identity. They keep relationships that do not depend on a single app. The shift is refusing to let any one platform become the gatekeeper of your life.

This chapter is about rejecting the false choice between reckless participation and total withdrawal. The constraint is that modern systems treat visibility as the default and punish boundaries through friction and social pressure. Women are expected to be reachable, searchable, and legible. The shift is staying engaged through design: separating audiences, controlling discoverability, choosing friction intentionally, and building exit routes that preserve autonomy.

The point is not to become invisible. The point is to become unexploitable at low effort. Women can stay connected without surrendering constant access. They can build a public voice without tying it to every personal detail. They can participate without accepting that exposure is the entry fee. Protection without withdrawal is not a contradiction. It is the only realistic posture for women who still need to live, work, speak, and move inside systems that were not built for their safety.

~8
Global Systems, Local Impact

The Constraint

Women are told the digital world is the same
everywhere. Same apps. Same platforms. Same phones.
Same rules. Same threats. The technology feels
universal, so women assume the risks are universal too.
But digital control is not experienced equally across
borders, communities, or households. The system is
global. The consequences are local. And the pressure
lands differently depending on where a woman lives
and what kinds of power surround her.

A global system is any digital infrastructure that reaches
across regions and standardizes behavior. Social
platforms. App stores. Cloud services. payment rails.
identity verification tools. messaging systems. search
engines. These systems operate at scale, which means
they are designed to treat people as categories, not
individuals. They cannot understand context at the
local level, so they enforce one set of assumptions across
many different realities. Women are expected to adapt
to the assumptions, even when the assumptions don't
match their lives.

Most of the control is not felt as censorship. It is felt as
dependence. A woman may rely on one messaging app
to communicate with family. One platform to promote
work. One phone number to access banking. One email

account to manage school and healthcare. The system feels convenient when it works. But when a global platform becomes the gatekeeper of local life, control becomes invisible. A woman's access depends on compliance with rules written far away, enforced automatically, and applied without nuance.

This matters because global systems often reflect the priorities of powerful institutions, not the needs of women. Privacy standards are set by corporate incentives. Moderation policies are shaped by public relations. Verification rules are shaped by fraud models and legal pressure. Account recovery is designed around stable identities and stable access. Women living in unstable conditions pay the price first. If a woman changes numbers, changes addresses, shares devices, or moves frequently, the system treats her as suspicious even when she is simply surviving.

Digital systems also travel with political climates. A platform may present itself as neutral, but it operates inside laws, regulations, and enforcement regimes that vary by country. In some places, surveillance is open. In others, it is quiet. In some places, speech is monitored. In others, access is restricted through informal pressure. Women are affected by these differences because women are often closer to social control, family control, and institutional control at the same time.

A woman's risk depends on what happens when she is exposed. In one environment, exposure might mean embarrassment. In another, it might mean job loss. In another, it might mean family punishment. In another, it might mean legal consequence. The same digital trace can be interpreted as harmless in one place and dangerous in another. Global systems do not adjust for that. They create records anyway. They store history anyway. They apply policy anyway. Women carry the consequences.

This is why "one-size-fits-all" digital safety advice fails. Advice is often written from the perspective of stable infrastructure. Stable banking. Stable legal protection. Stable access to phones and devices. Stable housing. Stable identity documents. But many women do not live inside stable conditions. They share devices. They use prepaid phones. They change SIM cards. They rely on public Wi-Fi. They coordinate through family accounts. They juggle multiple roles with limited access. Global systems treat those realities as anomalies instead of normal life.

Global platforms also change local culture. They shape what is considered normal behavior. They create new expectations around availability, responsiveness, and visibility. A woman may be pressured to maintain a public presence to be taken seriously. She may be pressured to share photos to prove legitimacy. She may be pressured to participate in group chats that blur boundaries. The platform's design becomes social enforcement. Women do not just adopt tools. They adopt norms that come with the tools.

This matters because women already live inside cultural systems that regulate them. Expectations about modesty, obedience, reputation, marriage, family roles, sexuality, and "proper" behavior. Digital life amplifies these pressures because it creates records that can be shared and preserved. A private conversation becomes a screenshot. A small comment becomes a rumor. A photo becomes evidence. Global platforms make those records easy to distribute, and local power structures use them as enforcement tools.

Women are also affected differently by financial infrastructure. In some places, women cannot easily open bank accounts. In other places, accounts exist but are monitored. In other places, payments require identity verification tied to family or state records. Digital money systems can become liberation, but they can also become control. A woman's ability to buy, move, or save can depend on permissions she does not fully own. A frozen account is not an inconvenience everywhere. In some environments, it is crisis.

Even when the technology is the same, the harm is shaped by local institutions. Police response. Legal protections. Employer behavior. School policy. Medical privacy. Cultural stigma. A woman's ability to report harassment depends on whether reporting is safe. A woman's ability to change identity depends on whether change is allowed. A woman's ability to leave depends on whether leaving is socially survivable. Digital exposure can trigger local consequences that global systems do not acknowledge.

Global systems also create language and identity problems. Many systems do not handle non-Western names well. They do not handle shared family devices well. They do not handle migration well. They do not handle women's name changes well. They do not handle lost documents well. Women are pushed into verification loops they cannot satisfy. Accounts are locked. Access is delayed. Services fail. The platform calls it security. The woman experiences it as being trapped outside her own life.

The constraint becomes deeper because global systems centralize records. Cloud storage. Messaging archives. Platform histories. Biometric identity checks. These systems create permanence. And permanence becomes leverage when a woman needs to change her life. A global record follows her across borders, employers, relationships, and institutions. The record does not care that context changed. It continues to exist as a searchable shadow.

This is the constraint: digital control is global, but its consequences are local. The same systems that make life easier can also act as gatekeepers, enforcement tools, and record-keepers that outlast context. Women experience those effects differently depending on culture, law, infrastructure, and social power. Global systems treat women as standard users. Local reality determines what exposure actually costs.

The Shift

The shift begins when women stop looking for one universal safety playbook and start building principles that adapt. Global systems are designed to standardize behavior, but women's lives are not standardized. Culture varies. Laws vary. Infrastructure varies. Family power varies. Economic dependency varies. The same app can be harmless in one place and dangerous in another. So the goal is not one perfect set of rules. The goal is portable strategy.

Women are often given digital advice that assumes stability. Stable phone numbers. Stable devices. Stable housing. Stable documents. Stable legal protections. Stable access to money. That advice is incomplete because it treats instability as rare. But instability is common for women. Women change numbers. Share devices. Move households. Live with relatives. Work in informal economies. Travel across borders. Rebuild after separation. The shift is designing for the reality that women's lives may change quickly, even when systems demand consistency.

This begins with one core idea: safety cannot depend on permission. If a woman's survival requires a platform to cooperate, she is fragile. If her access requires a customer support ticket, she is fragile. If she needs an institution to believe her quickly, she is fragile. This is not a moral judgment. It is a structural truth. Global systems are slow, automated, and optimized for scale. Women need strategies that remain functional even when institutions are indifferent.

The first practical shift is reducing single points of failure. Global platforms love centralization. One login across services. One cloud archive. One identity anchor. One phone number that unlocks everything. This structure is convenient until it becomes control. When women build redundancy, they loosen the grip of the system. A backup route matters more than a perfect tool. Another way to access money. Another way to communicate. Another way to recover accounts. Another place to store important files. Redundancy is what allows women to adapt without panic.

This is especially important in environments where documents and identity are contested. Women may not control household paperwork. Women may not control family accounts. Women may not have stable proof of address. Women may not be able to reset accounts without someone else's cooperation. The shift is recognizing that "official" identity is not always fully owned by the woman living inside it. So she builds practical independence where she can, in small layers.

Women also need a strategy for local surveillance. In some places, surveillance is institutional. In others, surveillance is social. Family networks. community pressure. employer monitoring. religious enforcement. neighborhood gossip. The technology is the same, but the watcher changes. Women protect themselves by controlling what is visible and by limiting what becomes permanent evidence. Not because they are hiding wrongdoing, but because they know that context can be weaponized.

This is where minimization becomes a global tool. The fewer systems that collect, the fewer systems that can expose. The fewer records that exist, the fewer records that can be used later. Women do not need to erase themselves from life. They need to reduce the permanent trail that outlasts context. Minimization travels well because it does not depend on a specific app or a specific government policy. It depends on a woman choosing smaller exposure as her baseline.

Another shift is separating categories of life so they do not collapse into one record. Women often operate across multiple roles at once: caregiver, worker, organizer, partner, daughter, community member. Global platforms merge these roles into one profile. That makes women easier to manage and easier to punish. Separation makes women harder to corner. One account is not the full story. One breach is not total collapse. One audience cannot reach every part of her life.

Women also adjust by choosing tools based on exit, not hype. The question is not "Is this popular?" The question is "Can I leave?" Can she export what matters? Can she recover without begging? Can she move to another tool without losing everything? Global platforms change quickly. They get pressured by governments. They get pressured by advertisers. They change moderation rules. They change search visibility. They shut accounts down. Women need platforms that do not trap them.

This is also where women stop trusting "safety" branding. Safety claims are often marketing. "Verified" does not mean protected. "Encrypted" does not mean safe in every context. "Private" does not mean not collected. "Secure" does not mean not weaponizable. Women do not need cynicism. They need clarity. Safety is not a label. Safety is what happens under pressure, when someone tries to use systems against you.

Women also need to understand that global systems export norms. They normalize constant reachability. They normalize public identity. They normalize oversharing. They normalize documentation. They normalize visibility as credibility. Those norms can be dangerous in cultures where women are punished for being visible in the wrong way. The shift is refusing imported norms that increase risk. A woman does not need to perform modern digital openness to be legitimate. She can choose privacy as competence.

Women also regain power by planning for instability without panic. Many women fear that preparing for worst-case scenarios makes them dramatic. It doesn't. It makes them resilient. A woman can quietly prepare for account lockout, device loss, financial restriction, or social retaliation without announcing it. She can treat preparation as routine maintenance. That is the shift: normalization of readiness.

This shift includes protecting what matters most: communication, money, identity, and movement. Those are the pillars that determine whether a woman can adapt when circumstances change. If she can communicate, she can coordinate support. If she can access money, she can act. If she can control identity pathways, she can recover. If she can move without being tracked, she can protect herself and her family. Women do not need to win every digital battle. They need to keep these pillars stable.

This chapter is about seeing global systems clearly. The constraint is that global platforms standardize behavior and create permanent records that outlast context, while local realities determine what exposure costs. Women face different consequences across cultures and infrastructures, but the same mechanisms appear everywhere: centralization, retention, and enforcement without nuance. The shift is building portable strategy: redundancy over dependence, minimization over exposure, separation over collapse, and exit over entrapment.

The point is not to become rigid. The point is to remain functional under changing conditions. Global systems will continue to expand. They will continue to collect. They will continue to pressure women into standard profiles and permanent records. Women who build portable strategies keep control even when policies shift, cultures tighten, or systems fail. That is what autonomy looks like across borders: not a single solution, but a set of principles that travel with her.

~9
Limits of Digital Protection

The Constraint

Women often start digital self-defense by looking for the right tool. The right app. The right settings. The right device. The right security checklist. It makes sense. Tools are tangible. They feel like control. They offer the promise that if you do everything correctly, you can be safe. But this is where many women get trapped. Digital protection is real, but it has limits. And the most dangerous mistake is believing there is a product that can substitute for power.

The first constraint is that systems are not designed around women's safety. They are designed around scale, compliance, and liability. A platform's goal is not to protect a woman's life. It is to retain users, reduce costs, and avoid public backlash. A bank's goal is not to preserve a woman's autonomy. It is to reduce fraud and meet regulations. A workplace's goal is not to maintain a woman's boundaries. It is to monitor productivity and protect the organization. Women are operating inside systems that may offer security features, but do not share women's priorities.

This is why protection often fails under pressure. When something goes wrong, women meet automation. Forms. tickets. templates. delays. A woman needs urgency. The system needs procedure. She needs context. The system needs proof. She needs a human response. The system gives her a policy. These are not personal failures. They are structural limits of large-scale digital governance.

The second constraint is that protection cannot eliminate the record. Even the best tools still produce traces. Devices generate metadata. Networks log connections. Platforms retain histories. Institutions store records. A woman can encrypt a message, but the fact that communication happened may still be visible. A woman can delete an account, but copies can exist elsewhere. A woman can lock down settings, but other people's screenshots and archives can still preserve what she tried to remove. The record is larger than any one tool.

Women also face the constraint of other people. Digital self-defense assumes the user controls her environment. But women often do not control the environment. A woman may share devices with children or family. She may be pressured by a partner. She may be monitored by an employer. She may be surrounded by people who overshare. A woman can tighten her own settings and still be exposed by someone else's defaults. No tool can fully protect a woman from the choices of the people around her.

There are also limits created by dependency. A woman cannot protect what she cannot leave. If her child's school requires a portal, she must use it. If her job requires monitoring software, she must comply or lose income. If her healthcare provider requires an app, she may have no alternative. If her bank requires phone-based verification, she may be forced to keep the number stable. Many women are locked into systems that demand exposure as the price of functioning.

This is why "perfect privacy" is not realistic. Most modern participation requires identity. It requires accounts. It requires verification. It requires logging. Systems are designed to be auditable because auditability is valuable to institutions. A woman can reduce exposure, but she cannot always remove it. And if she believes she can fully erase herself, she may take risks based on false confidence.

The constraint becomes sharper because digital protection is uneven. Some women can afford private devices, private plans, and secure storage. Some women can afford legal support, paid services, and reliable backups. Some women can take time off to rebuild accounts after a breach. Others cannot. Many women are managing safety while exhausted, under financial pressure, with children in their care, and limited time. Tools that require constant maintenance often fail in real life because women do not have the bandwidth to keep them perfect.

Women also face the constraint of misunderstanding. Safety is often marketed as a checklist. Turn on two-factor. Use a strong password. Update your apps. Those are good steps. But they do not address the deepest risks women face: coercion, stalking, harassment, institutional indifference, and social punishment. A woman can have strong passwords and still be pressured to reveal them. She can have encryption and still be doxxed. She can have private accounts and still be targeted through her children's devices. Protection fails when the threat is human and persistent.

Another hard limit is that systems can change without consent. A platform can rewrite policies. A device update can alter permissions. An app can introduce new collection. A service can be bought and change behavior. What was private can become searchable. What was contained can become shareable. Women are blamed for not keeping up, even though no one can perfectly track every shift. This is a structural limit of living inside digital ecosystems that women do not control.

The constraint also includes the psychological cost. Constant vigilance drains women. If protection requires a woman to monitor every interaction, every setting, every threat, she will eventually burn out. Burnout creates mistakes. Burnout creates shortcuts. Burnout creates resignation. That is not a weakness. It is human reality. Systems that require constant attention are not truly protective. They simply shift the labor onto women.

This is why the deepest limit is that digital safety cannot replace social safety. A woman can reduce exposure, but she cannot control how people respond to her boundaries. She cannot control whether institutions believe her. She cannot control whether a community punishes her. She cannot control whether someone escalates. Tools can help. But tools cannot eliminate power imbalances. Women need protection that recognizes these limits so they do not build false confidence.

This is the constraint: no tool can fully prevent harm in a world built on retention, surveillance, and enforcement through systems that are not designed for women's safety. Digital protection is real, but it is partial. Women can reduce risk, but they cannot buy immunity. The goal is not perfect security. The goal is durable options when the system fails and when people behave badly.

The Shift

The shift begins when women stop treating digital self-defense as a promise and start treating it as leverage. Tools do not create immunity. They create advantage. They reduce risk. They buy time. They preserve options. They help a woman keep functioning when something breaks. The moment women expect tools to guarantee safety, they end up betrayed by the first failure. The moment women treat tools as leverage, they can handle failure without collapse.

This matters because women are often sold the fantasy of "complete security." Buy this. Install that. Turn this on. Follow this checklist. Then you'll be safe. That framing is comforting, but it is dishonest. Modern life is built on retention, monitoring, and dependence. Institutions log. Platforms store. Systems enforce. And humans escalate when they want control. A woman cannot purchase her way out of those structures. She can only build for reality.

So the shift is moving from perfection to posture. Posture means a woman's safety does not depend on always doing everything right. It means her life does not collapse when she makes a mistake. It means one failure does not become total exposure. It means she can recover quickly. Posture is not obsessive vigilance. It is resilient design.

This starts with knowing what protection can and cannot do. A password manager can prevent password reuse, but it cannot prevent coercion. Two-factor authentication can reduce remote takeover, but it cannot stop a controlling partner who has physical access. Encryption can protect message content, but it cannot erase the fact that communication exists. Private accounts can reduce random harassment, but they cannot stop someone who already knows your identity. The shift is using tools for what they do best, without asking them to solve what they cannot.

Women also need to assume that systems will fail. Not because they are careless, but because failure is built into scale. Accounts get locked. Algorithms misread behavior. Support systems delay response. Devices break. Networks go down. Policies change. The shift is planning for failure as normal. A woman who plans for failure does not panic when it happens. She moves to her backup route.

This is why redundancy becomes the core strategy when tools have limits. One email should not be the only recovery path. One phone number should not unlock everything. One device should not be the only key to money and communication. One platform should not be the only place where work lives. Women do not need ten backups. They need one backup for the things that would collapse their life. Redundancy is how women keep autonomy when systems become gates.

Women also need to build safety that accounts for other people. Many women are exposed not because they made a technical mistake, but because someone around them overshared, pressured access, or refused boundaries. Tools cannot force other people to respect privacy. The shift is treating privacy as a relationship practice, not just a settings menu. Women can set expectations with friends, family, children, and colleagues about what gets posted, what gets tagged, what gets shared, and what gets stored.

This includes the harder truth: some environments will not respect boundaries. A woman may be in a workplace that monitors everything. A home where devices are shared. A family system that treats privacy as disobedience. A community that punishes women for being independent. The shift is recognizing that in these environments, the goal is harm reduction, not ideal safety. A woman focuses on protecting the most critical areas: money access, private communication, identity recovery, and movement. She protects what she must preserve to keep options open.

Women also shift by treating "privacy" as selective. Many women try to protect everything equally and burn out. The better strategy is priority. Some data categories matter more. Location. Financial access. Children's information. Private communications. The shift is giving those categories the highest protection and allowing lower-stakes areas to remain functional. This reduces decision fatigue and keeps protection sustainable.

Another shift is accepting that documentation is not the same as control. Women are told to save every screenshot, keep every log, collect every piece of proof. Sometimes that is necessary. But it also creates more material that can leak, be misused, or be weaponized later. The shift is documenting intentionally: only what matters, stored in a way that is controlled, and not mistaken for safety. Evidence helps in some systems. It does not prevent harm by itself.

Women also stop expecting platforms to protect them reliably. Platforms protect themselves. Their tools are imperfect, their incentives are mixed, and their responses are inconsistent. The shift is treating platform safety features as partial, not final. Blocking is useful, but it is not a shield. Reporting is useful, but it is not enforcement. Privacy settings are useful, but they do not eliminate data collection. A woman uses these tools, but she also builds safety beyond them.

This is where women reclaim power through exit. Exit is the only mechanism that platforms fully respect, because it threatens retention. A woman who can leave without losing everything holds leverage. She can move community. Move work. Move identity presence. Move communication channels. She can survive without one platform's approval. The shift is designing for exit: keep copies of important work, maintain independent contact routes, and avoid building life inside one platform's locked garden.

Women also benefit from strengthening their internal stance. Digital systems exploit panic. Harassers exploit emotional reaction. Institutions exploit exhaustion. The shift is recognizing that calm is a security advantage. Not because calm prevents harm, but because calm preserves judgment. A woman who can think clearly under pressure makes better moves. She does not hand over more data in a rush. She does not make irreversible decisions out of fear. She buys herself time.

This chapter is not meant to discourage protection. It is meant to keep women from building false confidence. The constraint is that tools have hard limits in a world built on retention, surveillance, dependency, and power imbalances. No system can guarantee safety for women. The shift is using tools as leverage, building redundancy for failure, prioritizing what matters most, and designing for exit and recovery instead of perfection.

The point is not to achieve "total security." The point is to remain uncornerable. A woman who knows the limits does not surrender when a system fails. She adapts. She uses backups. She reduces exposure where it counts. She refuses coerced access. She keeps options alive. Digital self-defense is not a magic wall. It is a structure that helps a woman keep moving when the world tries to trap her.

~10
Awareness as Defense

The Constraint

Women are often taught that digital safety is about fear. Be afraid of strangers. Be afraid of hackers. Be afraid of the internet. Be afraid of what might happen. The result is not protection. The result is paralysis. Women either avoid the topic entirely because it feels exhausting, or they spiral into constant vigilance that drains their life. Fear does not create durable safety. It creates stress and confusion. What women need is understanding.

Understanding does not mean knowing every technical detail. It means seeing structure. It means recognizing how systems work so a woman is not surprised when friction appears. It means noticing incentives so she can predict how a platform will behave. It means seeing where leverage is created so she can reduce it. Awareness is not paranoia. Awareness is orientation.

But most women are denied orientation. They are given slogans instead. "Use strong passwords." "Turn on two-factor." "Don't click suspicious links." Those are not wrong. They are just incomplete. They treat risk like a few isolated dangers instead of an environment. They make safety feel like a checklist instead of a posture. Women can follow every tip and still be harmed because the structure is still stacked against them.

The constraint is that digital systems are intentionally confusing. They are not built for clarity. They are built for compliance. The settings are buried. The defaults are permissive. The choices are framed to push women toward sharing. The language is softened so extraction feels normal. "Allow access." "Improve your experience." "Personalized content." "Recommended for you." These phrases hide what is actually happening: collection, retention, profiling, and behavioral shaping.

Women are also pressured to outsource understanding. "Just trust the platform." "Just trust the app." "Just trust the experts." In some areas, outsourcing is normal. You do not need to understand how an engine works to drive a car. But digital systems are not only tools. They are governance structures. They determine access. They store histories. They enforce rules through automated gates. Outsourcing awareness means outsourcing control. And women are the ones who pay when control is outsourced.

This is why women end up blindsided. A woman thinks an account is private, until it isn't. She thinks an app is safe, until a breach happens. She thinks a message is ephemeral, until it is screenshot. She thinks a platform protects her, until harassment escalates. She thinks a bank will understand context, until a transfer is frozen. She thinks a device is hers, until an update changes her settings. Each surprise is a reminder that she never owned the system. She only participated in it.

Awareness is also weakened by time pressure. Women make fast decisions because life demands speed.

Children need pickup. Work needs response. Family needs coordination. A woman clicks accept because she does not have time to investigate. She agrees because refusing means delays. She shares because it solves a short-term problem. This is not stupidity. It is survival under load. And systems exploit that load.

The constraint deepens because women are trained to doubt their own instincts. When a woman feels uneasy, she is told she is dramatic. When she wants boundaries, she is told she is paranoid. When she refuses access, she is told she is hiding something. This social pressure makes women second-guess themselves. It makes them hesitate to take protective action until something becomes severe. The system benefits from women delaying boundaries. The system benefits from women staying compliant.

Fear-based safety culture also creates mistakes. When women are afraid, they overcorrect. They either withdraw completely and lose access to opportunity and community, or they attempt drastic changes that collapse under real life. They try to lock down everything at once, get overwhelmed, and return to the easiest defaults. They download random "secure" apps, trust the wrong tools, and create more confusion. Fear makes women vulnerable because it reduces judgment.

Another constraint is that digital harm often looks small until it becomes serious. A little oversharing seems harmless. A small leak seems manageable. A few permissions seem normal. A little stalking feels like annoyance. A weird message feels like nothing. But harm is cumulative. Records persist. Patterns form.

Entitlement grows. Surveillance normalizes. And because women were never taught to see structure, they notice too late that a trail was built.

Women also face the constraint of competing responsibilities. Digital self-defense is often framed as a personal hobby. But for women, it is layered on top of caregiving, work, emotional labor, and financial management. The system demands constant attention, but women do not have infinite attention. When women are overloaded, they default to convenience. They ignore warnings. They keep using insecure settings. They stop updating. They stop tightening gaps. Not because they do not care, but because they are keeping life running.

Platforms also benefit when women stay unaware. Unaware women share more. Unaware women click more. Unaware women stay longer. Unaware women accept tracking. Unaware women provide richer profiles. Awareness disrupts that. Awareness makes women minimize. Awareness makes women separate audiences. Awareness makes women question defaults. That is why clarity is not prioritized. Confusion is profitable.

Confusion is also protective for institutions because it limits resistance. If women cannot clearly explain what is happening, they cannot argue for change. They cannot demand better defaults. They cannot negotiate boundaries at work or at home. They cannot push back against policies that violate privacy while claiming "security." Uncertainty keeps women compliant because compliance is easier than debate.

This is why the system keeps moving even when women are uncomfortable. A woman senses risk, but she cannot name it cleanly. She sees friction, but she cannot trace it to design. She feels exposed, but she cannot locate where the exposure is coming from. Without clarity, she defaults to endurance. She adapts. She tolerates what should not be normal. That is how digital environments shape women through exhaustion instead of consent.

The most dangerous part is that women are often told digital safety is an individual responsibility. If a woman is harmed, she is asked what she did wrong. Why did she post that? Why did she answer? Why did she share? Why didn't she block? Why didn't she report? The burden is placed on women to anticipate every threat and manage every escalation. Meanwhile, the systems that created the exposure remain framed as neutral. Women carry the blame for an environment designed to extract and expose.

This is the constraint: women are pushed into a fear-based relationship with technology instead of an informed relationship. They are kept busy, pressured to comply, and shamed for setting boundaries. Systems are designed to obscure incentives and normalize tracking, retention, and dependency. Without awareness, women are surprised by the cost of participation. And when women are surprised, they lose time, autonomy, and control at the exact moments they need them most.

The Shift

The shift begins when women stop treating digital
safety as a mood and start treating it as literacy. Fear is
not a strategy. Fear is a reaction. It spikes attention for a
moment and then burns women out. Awareness does
the opposite. Awareness steadies a woman's decisions
over time. It gives her language for what she is seeing,
which means she is less likely to freeze, comply
automatically, or spiral into exhaustion when something
feels off.

This is the first change: women stop trying to become
"perfectly secure" and start trying to become harder to
mislead. Most harm begins with confusion. A woman
clicks because the prompt is urgent. She agrees because
the system framed refusal as impossible. She shares
because she was rushed. She stays exposed because she
cannot tell which settings matter and which ones are
just noise. Awareness reduces these openings because it
restores a woman's ability to pause and interpret.

Awareness starts with recognizing incentives. Platforms
are not neutral. They are businesses. Their profit comes
from attention, data, and retention. That means they
will always encourage more sharing, more linking, more
syncing, more public defaults, and more time spent
inside the feed. A woman does not need to hate the
platform to understand what it wants. She only needs to
stop confusing a company's incentives with her own
needs.

Once a woman understands incentives, she can predict behavior. She can predict that default settings will drift toward visibility. She can predict that new features will increase collection. She can predict that "personalization" will deepen profiling. She can predict that safety tools will often protect the platform first. That prediction matters because it turns surprise into expectation. When women expect drift, they are less likely to be caught off guard.

The shift also includes learning the difference between convenience and control in real time. Convenience feels good immediately. Control feels boring. Control feels like extra steps. Control feels like saying no to default access. But control is what prevents collapse later. Awareness helps women choose the right trade at the right moment. A woman does not reject convenience entirely. She uses it consciously, and she stops letting it become dependency by accident.

This is where women stop thinking about safety as a one-time setup. Digital life is not static. Platforms change. Policies change. Tools update. Your life changes. Your risk changes. Awareness turns safety into maintenance instead of obsession. It means women do not have to constantly panic-check everything. They just need a consistent posture: reduce what is collected, reduce what is stored, reduce what is linked, and keep exit routes intact.

Awareness also gives women language for boundaries. Boundaries are harder to hold when a woman cannot explain them. People exploit vagueness. They push until she gives in. "Why not just share your location?"

"Why are you being weird about your phone?" "Why can't I see your messages?" These demands are often framed as trust. Awareness gives women a clean response: privacy is normal. Access is not owed. Transparency is not the price of peace. A woman does not need to argue every detail. She only needs to refuse the entitlement.

Women also learn to see digital harms before they become emergencies. Awareness means recognizing early signals: a platform encouraging a real-name link, a client app requesting full contacts, a new feature making an old archive searchable, a friend casually tagging your location, a workplace portal expanding monitoring, a partner escalating access requests. None of these moments are dramatic. They are the start of a pattern. Awareness helps women act early while the cost is still small.

This is why awareness is a form of time management. Women lose the most time when they wait until something breaks. Account recovery, harassment escalation, identity lockouts, device replacement, and damage control can cost days. Awareness costs minutes. A quick decision to deny unnecessary access can prevent hours of cleanup later. A quick decision to separate identities can prevent a full cascade later. Awareness is a way to buy back time.

Awareness also reduces shame. Many women carry shame around past choices. They feel stupid for sharing. They feel guilty for trusting. They feel embarrassed for not knowing. Shame is useless. Shame keeps women quiet and compliant. Awareness replaces shame with clarity. A woman can accept that systems are designed to be confusing and still take control now. The goal is not to punish her past self. The goal is to protect her future self.

This is where women stop treating their digital life like a single space and start treating it like an ecosystem. Different spaces require different levels of exposure. Some spaces are public. Some are semi-public. Some are private. Some are high-risk. Some are stable. Awareness lets women match their posture to the environment. They can be public without being personally accessible. They can be private without disappearing. They can stay engaged without being reckless.

Women also regain autonomy by recognizing that safety is not only technical. It is social. It is financial. It is structural. A woman's strongest defense is often not an app. It is redundancy. It is independent access to money. It is a backup communication route. It is a relationship that respects boundaries. It is a community that does not leak details casually. Awareness keeps women from being trapped in the fantasy that one tool can substitute for a stable foundation.

Digital self-defense is not about achieving perfect security. It is about preserving autonomy under pressure. Systems will continue to collect and retain. Platforms will continue to profit from exposure. Institutions will continue to enforce through automated gates. Women do not win by trying to control everything. Women win by staying oriented: minimizing what becomes permanent, separating what should not collapse together, building exit routes, and holding boundaries without apology.

Awareness is not a mindset exercise. It is a practical weapon. A woman who understands the system is harder to trap. Harder to rush. Harder to corner. She can participate without surrendering herself. She can use technology without being used by it. She can keep her life moving even when the environment is built for extraction.

This chapter is about using understanding as defense because understanding scales. Fear does not scale. Fear burns women out. Fear isolates them. Awareness scales because it turns into instinct over time. It makes women quicker at spotting risk patterns. It makes women calmer when friction appears. It makes women less vulnerable to manipulation through urgency and shame. It gives women a posture that holds up even when tools fail.

The point is not to make women afraid of the digital world. The point is to make women fluent inside it. Fear tells women to shrink. Awareness tells women to choose. And choice is the foundation of digital self-defense: not perfect safety, but durable control when pressure arrives.

Conclusion

Digital self-defense is not about disappearing. It is about staying present without being captured. It is the decision to move through modern life with boundaries intact, instead of treating exposure as the price of participation. A woman does not need to abandon technology to protect herself. She needs to stop letting technology define what she must surrender.

Self-defense starts with clarity. Clarity about what systems collect. Clarity about what gets stored. Clarity about what becomes permanent. Clarity about where control is concentrated and where it can be reclaimed. The goal is not to fear every tool. The goal is to see the tradeoffs clearly enough to choose on purpose.

This book has shown that protection is not one perfect setup. It is a posture built over time: reducing what accumulates, separating what should not collapse together, building redundancy, and keeping exit routes open. A woman becomes harder to trap when she is less legible by default and less dependent on one fragile path. She does not need total safety. She needs durable options when pressure arrives.

The point is not isolation. The point is autonomy. A woman can stay engaged, build a life, speak, create, work, and connect without surrendering full access to systems that were never designed for her protection. Digital self-defense is not paranoia. It is competence. It is clarity practiced daily until it becomes freedom.

About the author

The author lives removed.

Please feel free to burn part or all of this book, safely, as an effigy.

www.ingramcontent.com/pod-product-compliance
Lightning Source LLC
Chambersburg PA
CBHW020943090426
42736CB00010B/1246